ESSENTI

global warming

FRED PEARCE

SERIES EDITOR JOHN GRIBBIN

LONDON, NEW YORK, MUNICH,
MELBOURNE, DELHI

senior editor Peter Frances
senior art editor Vanessa Hamilton
US editor Eileen Nester Ramchardran
DTP designer Rajen Shah
illustrator Richard Tibbitts
category publisher Jonathan Metcalf
managing art editor Phil Ormerod

Produced for Dorling Kindersley by
Grant Laing Partnership
48, Brockwell Park Gardens, London SE24 9BJ

managing editor Jane Laing
editor Jane Simmonds
managing art editor Christine Lacey
picture researchers Jo Walton, Louise Thomas
indexer Dorothy Frame

First American Edition, 2002
02 03 04 05 10 9 8 7 6 5 4 3 2 1

Published in the United States by
DK Publishing, Inc., 95 Madison Avenue
New York, New York 10016

A Cataloging-in-Publication record for this title is
available from the Library of Congress
ISBN 0-7894-8419-6

Color reproduction by Mullis Morgan, UK
Printed in Italy by Graphicom

See our complete product line at **www.dk.com**

contents

the world is warming

Global warming is with us. A decade ago, it was a matter of conjecture. Now the future is unfolding before our eyes. The Inuit of North America see it in disappearing ice, starving polar bears, and wayward whale migrations. The people of shanty towns from Latin America to Southeast Asia see it in lethal hurricanes, landslides, and floodwaters. Europeans see it in vanishing alpine glaciers, Mediterranean droughts, and freak storms. Researchers see it in everything from tree rings and lake sediments to ancient coral and bubbles trapped in ice cores. All reveal that the world has not been warmer for a millennium or more. And it has probably never warmed so fast as it has in the past 25 years – a period when natural influences on global temperatures such as sunspots should, if anything, have been cooling us down. With the physics of the greenhouse effect a matter of scientific fact for more than a century, it is hard to disagree with the overwhelming majority of climatologists who say that what we are seeing is manmade climate change.

melting ice

The effects of rising temperatures can be seen most dramatically at the poles, where sea ice is melting earlier and refreezing later each year, and where its overall mass is decreasing.

the evidence

The warmest year in the warmest decade in the warmest century of the last millennium was 1998. The second- and third-warmest years on record were 1997 and 1995. There is now little doubt that our planet warmed substantially during the late 20th century – by 0.5°F (0.25°C) a decade, according to the US National Climate Data Center.

cracking up
This building in Yakutsk, in Russia's Siberian Arctic, is subsiding due to permafrost melting beneath it. If underground ice melts during summer, the soil may not have enough strength to support structures built on it.

Warming in the late 20th century was greatest in the Arctic. Much of Siberia has warmed by 9°F (5°C) – eight times faster than the global average – causing melting of permafrost, buckling of roads, and toppling of buildings. In Alaska, as the permafrost proves not to be so permanent, ice cellars used by hunters for centuries to store caribou meat and whale blubber in summer are melting.

global thaw

The Greenland ice cap is rapidly losing thickness on coastal margins. Sonar measurements by US and British military submarines reveal a decline in the average thickness of Arctic ice in late summer of 42 percent in the past 40 years. Ships can now sail through the legendary Northwest Passage above Canada most summers.

The evidence across most of the Arctic is of a rate and extent of warming beyond anything recorded before. Off Antarctica, krill find that the sea ice under which they feed is disappearing. Other creatures go hungry as a result. This may explain population crashes among the sea lions of the Falklands and elephant seals of the South Shetland Islands.

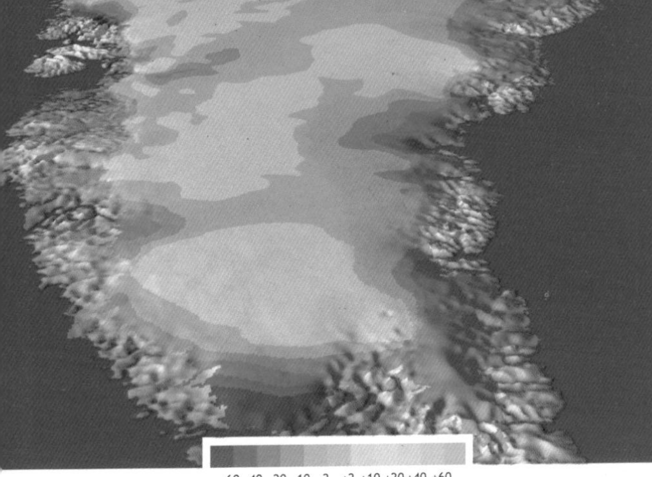

rate of change in ice cap height (cm/year)

greenland melts
The lower slopes of the great Greenland ice cap are melting fast, by over 18in (50cm) a year in places. The melting is overtaking the rate of accumulation of ice in the ice cap's heart, caused by increased snowfall.

The evidence of a widespread thaw extends far beyond the Arctic and Antarctic. Snow cover on the Earth's land surface has diminished by ten percent since the 1960s. The annual period during which lakes and rivers ice over has diminished by about two weeks. Mountain glaciers have retreated from peaks worldwide, including Mount Kilimanjaro in Tanzania, which has lost 82 percent of its ice cap since 1912.

"Last year, the ice stayed over the horizon all summer. We had to go over 30 miles to hunt seals. There's been a lot of climate change."

Eugene Brower, Barrow Whaling Captain's Association, Alaska, 2000

Alpine glaciers contain half as much snow and ice as they did when mountaineers first scaled their peaks in the mid-19th century. The Gruben glacier in Switzerland retreated by 650ft (200m) in the 1990s alone. The bodies of four British airmen, deep-frozen in northern Iceland since a plane crash in 1941, emerged in 1990. Only in Scandinavia are glaciers still growing – possibly because shifting storm tracks are dumping more snow on the area, compensating for any melting.

a long wait
In the Canadian north, polar bears are going hungry as they wait on land for the ice to re-form in fall and allow them back onto their marine hunting grounds.

scientific panel

Following an intense drought in the United States in 1988, the UN formed a scientific panel to investigate concern about global warming and what to do about it. The Intergovernmental Panel on Climate Change (IPCC) now involves more than 3,000 scientists and is the main advisory body to the UN's Climate Change Convention. Its third assessment, published in 2001, is the scientific basis for much of this book.

disappearing land

As ice on land melted and the oceans gained volume as a result of thermal expansion, sea levels worldwide rose by 4–8in (10–20cm) in the 20th century. Thermal expansion results from the warmth of the atmosphere penetrating into the oceans. Such warming has been detected to a depth of 10,000ft (3,000m) in the world's largest oceans, but is most marked in the top 1,000ft (300m), which have warmed by about 0.5°F (0.25°C) in the past 40 years. It would have been worse but for the storage of more water on land behind new dams. Four out of five of the world's beaches are eroding. Louisiana is losing 1 acre (0.4ha) of land every 24 minutes. First to go are barrier islands that protect the coastline from storm surges – responsible for 90 percent of deaths during hurricanes.

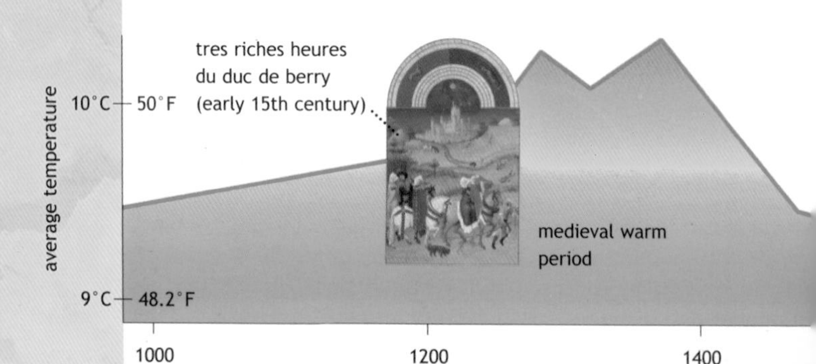

a natural phenomenon?

The weather is always changing, of course. Records are always being broken. Climate waxes and wanes. Nature is in a constant state of flux. How can we be sure that the current warming isn't simply part of a global cycle?

Temperatures have gone up and down to a considerable degree during the past 160,000 years. In this period there have been two ice ages. We now know that most of the northern hemisphere enjoyed a relatively warm period in the Middle Ages from the 12th to 14th centuries, followed by a "little ice age" that lasted fitfully until the early 19th century. Cultural evidence supports this: paintings of Dutch ice scenes by Bruegel and reports of

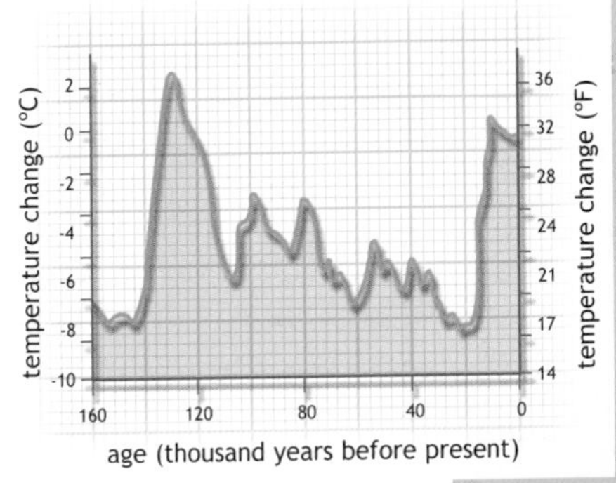

fluctuations
Readings from ice cores show temperature changes in Antarctica over the last 160,000 years, giving a context for current trends. The difference now is the speed of change.

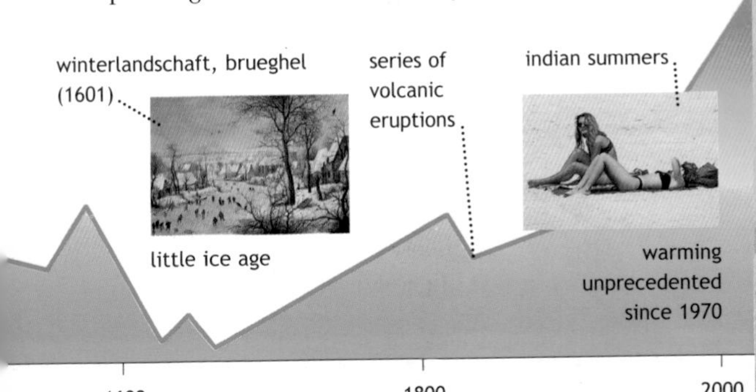

temperature change (°C)
2
0
-2
-4
-6
-8
-10
temperature change (°F)
36
32
28
24
21
17
14
160
120
80
40
0
age (thousand years before present)

temperature timeline
The last 1,000 years have seen significant rises and falls in global temperatures, but there has been nothing like the soaring rate of temperature change of recent years.

winterlandschaft, brueghel (1601)
little ice age
series of volcanic eruptions
indian summers
warming unprecedented since 1970
1600
1800
2000

measuring past conditions

Records of past temperatures and the proportions of gases in the atmosphere cover only the last few hundred years. Scientists can create longer records by studying natural phenomena that change due to temperature and atmospheric changes. These records are known as proxy records. The features studied include tree rings and air bubbles trapped inside ice cores.

tree rings

Each tree ring equates to a tree's annual growth – the thicker the ring, the faster the growth that year. Scientists can use these rings to determine the faster-growing (and therefore warmer) years.

ice cores

Analysis of air bubbles inside ice cores, and of coral and ancient marine organisms, reveals changing carbon dioxide levels in the atmosphere.

ice fairs on London's Thames River date from this period. Part of the warming and melting recorded since then can be put down to a recovery from that period. But in its latest report, the IPCC concludes that "the rate and duration of warming . . . in the 20th century cannot simply be considered a recovery." The Earth's surface is now warmer than at any time in the past thousand years.

natural causes of fluctuations

Scientists now claim to know quite a lot about the causes of past temperature fluctuations, whether they lasted for millennia or just a few years. Volcanic eruptions can cool the planet for months or even years by filling the upper atmosphere with a shading aerosol of particles. Changes in the amount of radiation leaving the Sun, shown for instance in sunspot cycles (see p.13), influence temperatures on Earth on timescales from a few years upward. They may have been responsible for the little ice age (see p.9) and some of the warming that followed it, starting in the 19th century. And subtle changes to the Earth's orbit known as the Milankovitch "wobbles" (see p.12), which change either the amount of heat reaching the planet's surface or its distribution, operate over thousands of years.

But natural causes cannot explain the recent temperature surge. Orbital wobbles should be propelling us toward the next ice age – scientists in the mid-20th century issued warnings of a cooler era coming. Recent volcanic eruptions have short-term effects, but do not affect long-term trends.

cooling volcanoes
When Mount Pinatubo erupted in the Philippines in 1991, the debris left in the atmosphere caused penetration of solar radiation to the Earth's surface to fall by two percent, resulting in cooling for two years.

milankovitch wobbles

The Serbian mathematician Milutin Milankovitch studied three kinds of "wobbles" in the Earth's orbit. They slightly affect the amount and, more considerably, the distribution of solar radiation that reaches the Earth, and so may slowly alter climate on long, timescales. In polar regions, these cycles can vary the amount of summer sun by ten percent. The changes in solar radiation may be amplified by "feedbacks" on Earth (see p.20), which affect the amount of radiation reflected back into space.

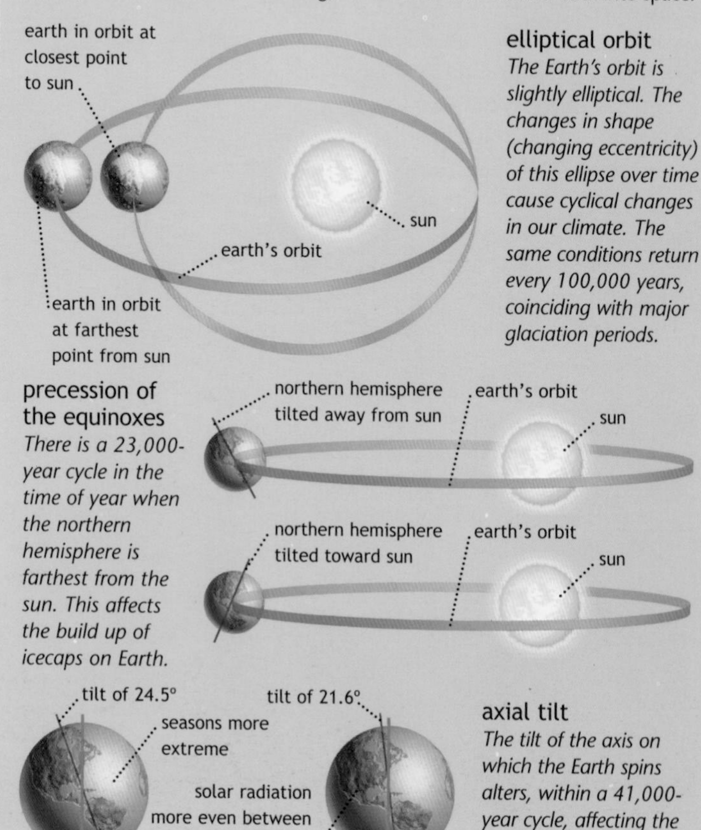

elliptical orbit

The Earth's orbit is slightly elliptical. The changes in shape (changing eccentricity) of this ellipse over time cause cyclical changes in our climate. The same conditions return every 100,000 years, coinciding with major glaciation periods.

precession of the equinoxes

There is a 23,000-year cycle in the time of year when the northern hemisphere is farthest from the sun. This affects the build up of icecaps on Earth.

axial tilt

The tilt of the axis on which the Earth spins alters, within a 41,000-year cycle, affecting the severity of the seasons.

The theory that solar cycles could explain variations in global warming gained credence during the 1990s, when Knud Lassen of the Danish Meteorological Institute reported in 1991 that sunspot cycles (see below) and rises in global temperatures seemed to have been coinciding over the previous century. But in 2000, Lassen announced that his latest analysis showed that solar cycles could not explain warming since 1960. "The curves diverge after 1960, and it's a startling deviation. Something else is acting on climate. It has the fingerprints of the greenhouse effect," he said.

This echoed the conclusion of the IPCC in its most recent reports that during the last quarter century, solar cycles should have been pushing us into colder times, when in fact precisely the opposite was occurring. So the finger points at human activity.

How could human activity be influencing temperatures on Earth? Scientists think it is due to a mechanism known as the greenhouse effect.

solar influence

Until around 1960, average global temperatures seemed to follow sunspot cycles. But since then the link has broken – implicating human activities.

temperature change (°C): 0.6, 0.4, 0.2, 0, -0.2, -0.4, -0.6
temperature change (°F): 33, 32.6, 32.3, 32, 31.6, 31.3, 31
year: 1850, 1900, 1950, 2000

actual temperature
predicted temperature (following sunspot cycles)

sunspots

Sunspots siginfy eruptions of activity on the sun. Intense magnetic fields form, and there is an increase in the radiation released toward Earth, where temperatures may rise slightly. The sunspots on the far left correspond to magnetic fields (blue and pink areas) in the image on the left.

sunspots

photograph taken in white light

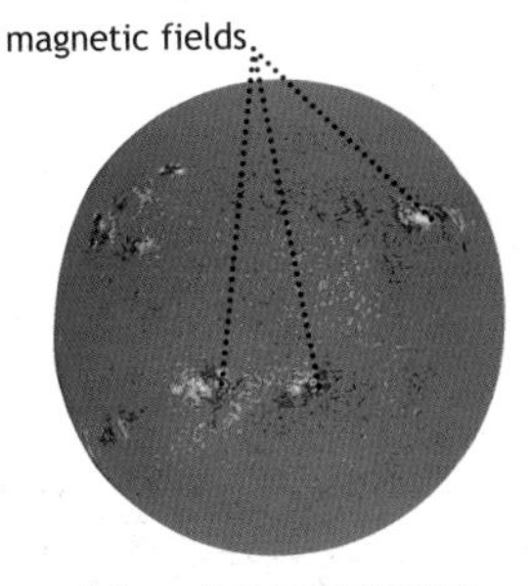

false color magnetogram

the greenhouse effect

Most scientists believe that the greenhouse effect is responsible for recent global warming. A natural greenhouse effect is essential for life on Earth. Physicists have known for nearly 200 years that certain gases in the atmosphere trap heat radiated from the Earth's surface, preventing it from escaping back into space (see panel, right). The key "greenhouse gases" for maintaining the Earth's equable temperature are water vapor (see p.20) and carbon dioxide (CO_2). Without their warming blanket, the Earth's surface would be frozen.

sweltering

Certain gases trap heat in the Earth's atmosphere, rather like the glass in a greenhouse – hence the term "greenhouse effect".

Each greenhouse gas has its own radiation "signature." Water vapor absorbs radiation with wavelengths between 4 and 7 micrometers; carbon dioxide absorbs wavelengths between 13 and 19 micrometers. The strengthening greenhouse effect of these gases was measured directly for

A French mathematician, **Jean Baptiste Fourier** (1768–1830), noted for his research on the theory of heat, was the first to use the greenhouse image for the way heat is trapped within the atmosphere. He argued in 1827 that the atmosphere acts "like the glass of a hothouse." Then, in 1860, Irish scientist **John Tyndall** (1820–93) measured the absorption of infrared radiation by carbon dioxide and water vapor, the most important of what came to be known as the greenhouse gases.

how the greenhouse effect works

Warmth from the sun heats the Earth's surface, which in turn radiates energy outward. Some of this outgoing heat escapes into space, but some is trapped in the atmosphere by the gases known as greenhouse gases. The radiation absorbed by the greenhouse gases heats the lower atmosphere, the troposphere. In fact, without these gases, the temperature of the Earth's surface would fall to below freezing. However, the speed of increase of these gases in the atmosphere is currently greater than ever before due to pollution from sources such as burning forests, vehicles (especially airplanes), industrial plants, and power stations.

some solar radiation penetrates atmosphere and heats earth's surface
some solar radiation bounces back into space
earth's surface radiates heat back into air
destroying forests releases CO_2 into the atmosphere
pollution causes CO_2 to accumulate in atmosphere, trapping more heat
greenhouse gases such as CO_2 absorb and so trap some of radiated heat

svante arrhenius

In 1894, Swedish chemist Svante Arrhenius calculated how much industrialization was adding to levels of critical gases in the atmosphere. In 1896, he wrote that if the amount of carbon dioxide in the air were to double, it would raise temperatures by 9–11°F (5–6°C) – a total close to today's estimates.

the first time in 2000, when scientists reported on changes in the spectrum of radiation escaping from the atmosphere into space. By comparing data collected by two satellites 27 years apart, they showed that less radiation is escaping to space from the frequencies that coincide with those of the main manmade greenhouse gases.

carbon dioxide

The amount of carbon dioxide in the air is now known to have closely shadowed global temperatures over hundreds of millions of years. It is this gas that humans are adding to the atmosphere in volumes sufficient to trigger climate change. Carbon dioxide is constantly exchanged between the atmosphere and the oceans, and is absorbed and released by plants and animals on the Earth's surface. Its

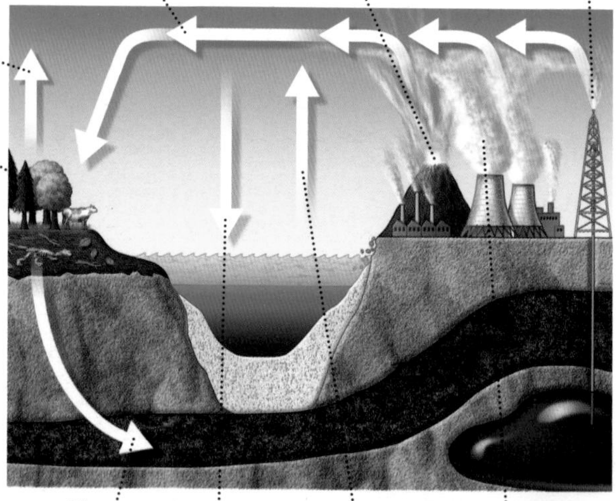
total release and absorption of CO_2 forms cycle
CO_2 released by natural events such as eruptions
oil brought from underground and burned, releasing CO_2
plants and animals release CO_2
forests and soil absorb CO_2 from the air
some rotting vegetation turns to fossil fuels, such as coal
ocean absorbs CO_2 from air
organisms in ocean release CO_2 into air
manmade sources release CO_2

carbon cycle
Carbon is naturally exchanged among the oceans, forests, and the atmosphere. Volcanoes add to the carbon in the system, while swamps where carbon becomes fossilized take it away. Humans have disturbed this natural balance.

concentration has a seasonal cycle, caused by its absorption by growing plants each spring and release each fall.

In the past 200 years, this seasonal cycle has been overlain by a long-term rising trend that sees each annual peak higher than the last. Initially, this rise was caused mainly by the destruction of forests, and the release of their carbon into the atmosphere. In the past 50 years or so, the dominant cause has been the burning of carbon-based fossil fuels such as coal and oil.

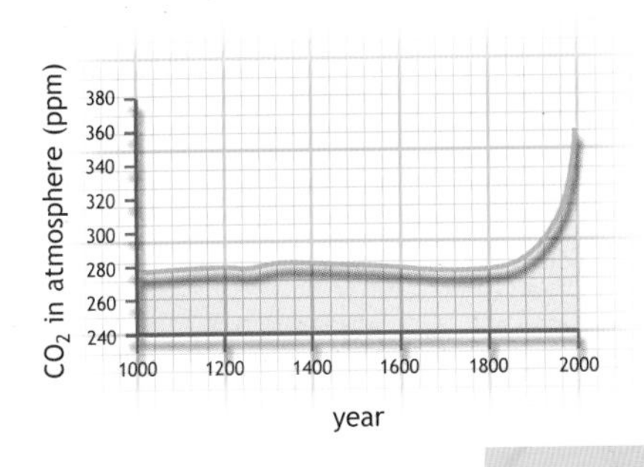

carbon dioxide levels
This graph shows how levels of carbon dioxide in the atmosphere have been rising swiftly since the 1800s from a previously steady level.

Up to half of the emissions are reabsorbed by the oceans and by plants, which grow faster in the carbon-dioxide-rich air. But there is an accumulation that causes an annual rise of around 0.4 percent in the gas's concentration in the air. Since 1800, the atmospheric concentration of carbon dioxide has risen from about 270 parts per million (ppm) to today's 370 ppm – a level thought to be higher than at any time in the past 20 million years.

methane

The second most significant greenhouse gas released by human activity is methane, which, molecule for molecule, is 20 times more potent as a greenhouse gas than carbon dioxide. It is produced largely through the actions of certain bacteria that have thrived in association with humans. These bacteria are found in the guts of ruminant animals, in landfills, and in rice paddies, for instance.

methane
Domesticated farm animals, such as cows and sheep, waste dumps, and farmland are now among the major sources of methane in the air.

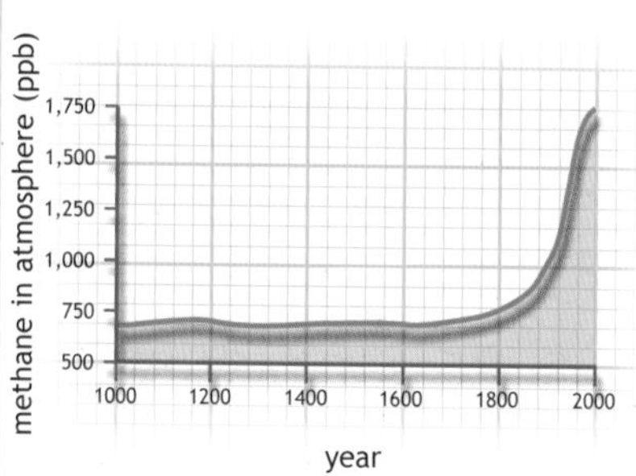

There are also releases of methane formerly trapped underground – from natural-gas pipelines and coal mines. Rotting vegetation in stagnant reservoirs is another significant source. One recent study suggested that reservoirs could account for up to one fifth of all methane emissions globally – causing seven percent of total global warming. Concentrations of the gas are now estimated to be higher than for at least 420,000 years.

methane levels
Measured in parts per billion (ppb), methane levels, like those of carbon dioxide, have risen sharply since 1800.

other greenhouse gases

Other manmade greenhouse gases that add marginally to global warming include nitrous oxide and ozone, which are also produced naturally, and some manmade compounds, such as chlorofluorocarbons (CFCs). CFCs have been largely phased out over the last 15 years to protect the ozone layer. But some of their replacements, such as hydrofluorocarbons, perfluorocarbons, and sulfur hexafluoride, used in refrigerators and other appliances, also add to global warming. Together, they produce three percent of Europe's contribution to global warming.

nitrous oxide levels
Nitrous oxide levels have been rising slowly for over a thousand years, but increasingly rapidly since 1800.

"The question is no longer whether the climate is changing in response to human activities, but rather how much, how fast, and where."

Bob Watson, chairman of IPCC, 2000

the ozone layer

Part of the stratosphere (upper atmosphere), the ozone layer shields the Earth's surface from harmful ultraviolet (UV) radiation from the sun. Some manmade chemicals containing chlorine and bromine float up into the stratosphere, where they destroy ozone. A thinner ozone layer lets more UV radiation into the troposphere (lower atmosphere), which has a slight warming effect. Some of these chemicals are also greenhouse gases, adding to the warming. This compunds damage to the ozone layer: by trapping heat near the Earth's surface, greenhouse gases cool the stratosphere, creating ideal conditions for more ozone destruction.

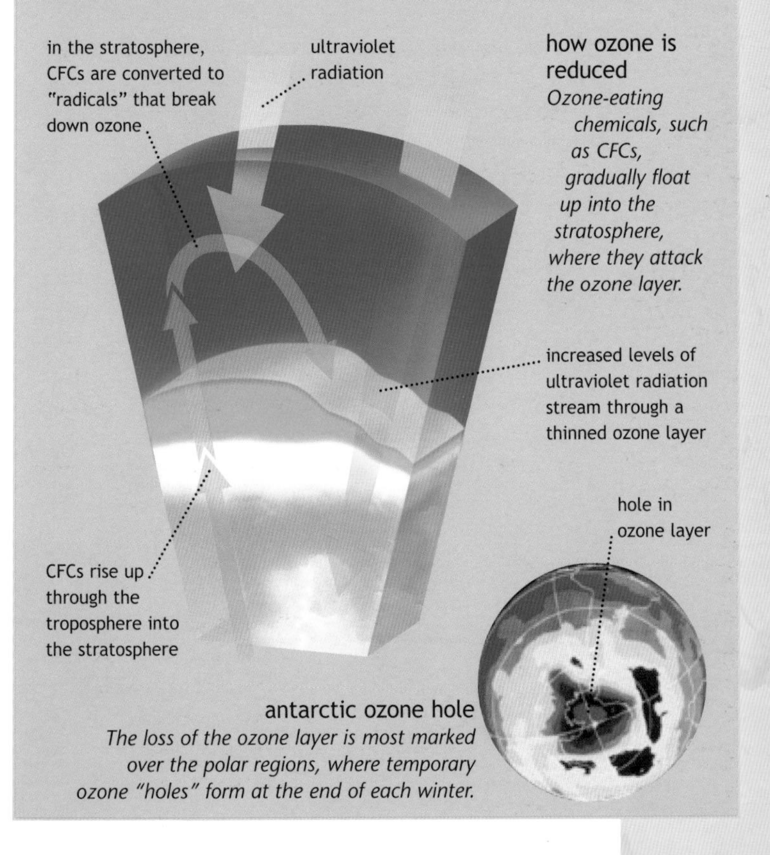
in the stratosphere, CFCs are converted to "radicals" that break down ozone
ultraviolet radiation
increased levels of ultraviolet radiation stream through a thinned ozone layer
CFCs rise up through the troposphere into the stratosphere
hole in ozone layer

how ozone is reduced

Ozone-eating chemicals, such as CFCs, gradually float up into the stratosphere, where they attack the ozone layer.

antarctic ozone hole

The loss of the ozone layer is most marked over the polar regions, where temporary ozone "holes" form at the end of each winter.

the impact of feedbacks

Scientists calculate that the pure physical effect of doubling the amount of carbon dioxide in the atmosphere from pre-industrial times – likely during the 21st century – may add 1.8°F (1°C) to global temperatures. But they expect that the real increase will be several times greater because of a series of "feedbacks," triggered by global warming, that could accelerate the warming. Positive feedbacks increase the temperature; negative ones have a cooling effect.

water vapor

The first positive feedback involves water vapor. A warmer world is likely to have more water vapor in the atmosphere, because the heat will evaporate more

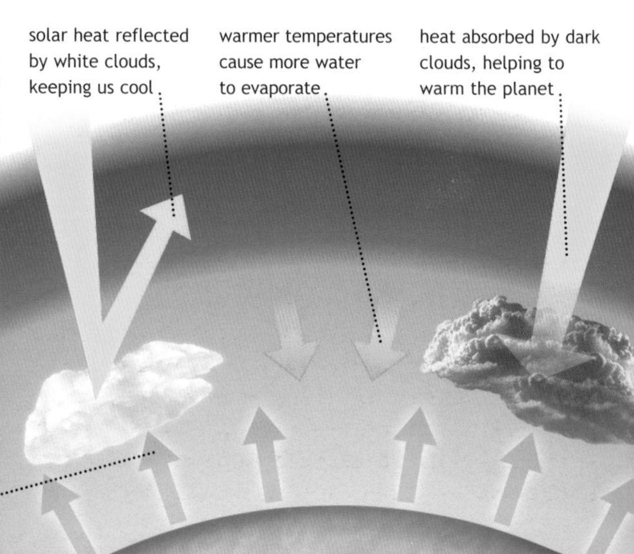
solar heat reflected by white clouds, keeping us cool
warmer temperatures cause more water to evaporate
heat absorbed by dark clouds, helping to warm the planet
increased evaporation forms more clouds, both white and dark

water vapor feedback
Water vapor is a powerful greenhouse gas, but it also forms clouds. Some clouds shade the Earth during the day, but others absorb solar heat and act as a blanket at night.

moisture, especially from the oceans. Water vapor is a greenhouse gas, and this positive feedback is likely to double the warming from carbon dioxide alone. But there are complications. Some water vapor will turn into clouds, and the effects of clouds are uncertain. They tend to cool the planet's surface, shielding it from the sun's rays during the day, but they also trap heat radiated from the surface, helping to keep the air warm, especially at night. The dominant effect depends on many factors, including the type and height of the clouds.

clouds
Clouds present one of the greatest uncertainties in calculating how much the greenhouse effect will warm the atmosphere.

snow and ice

A further major feedback is caused by snow and ice, which currently cover large areas of the Arctic and Antarctic. They reflect 80 percent of the solar radiation hitting the surface back into space, helping to maintain cold temperatures locally, and having a significant cooling effect on the planet as a whole. But as ice melts, the surface of the Earth exposed to the sun's rays will invariably be darker. Ocean water in particular is much darker. It absorbs 80 percent of the radiation. So melting ice will act – and is acting – as a positive feedback to warming, both locally and globally.

oceans

The oceans tend to slow warming by absorbing heat from the surface and carrying it to the ocean depths. The process is limited by the rate at which ocean waters move between different depths (vertical mixing), and oceanographers predict that global warming is likely to reduce the amount of mixing. So while oceans moderate global warming now, this beneficial effect could slow down in the future.

pollution

Dust, smoke, soot, and the tiny particles of sulfate that make up acid rain all shade the planet, taking some of the edge off global warming by reflecting radiation into space. The precise cooling effects of these particles are hard to assess, however – partly because some of them also absorb solar radiation and may re-radiate it rather as the ground does (see p.15). Current estimates are that the polluting particles, collectively called aerosols, may counteract as much as a quarter of the warming effect of greenhouse gases globally. In very polluted areas, they may entirely counteract it in the short term. But most aerosols last in the atmosphere for only a few days, whereas carbon dioxide persists for many decades, accumulating in the atmosphere.

the role of hydroxyl

One of the uncertainties in climate change is the role of a rare chemical in the air called the hydroxyl radical. Made up of one atom of oxygen and one atom of hydrogen, hydroxyl acts as the atmosphere's detergent. It reacts with a wide range of pollutants, including some greenhouse gases, removing them from the air. The fear is that ever-rising levels of pollution – especially in the tropics, where most hydroxyl is created naturally – will cause a shortage of the vital compound. If that happened, every pollutant would hang around in the air longer, causing worsening smogs and accelerating global warming.

In mid-2001, scientists at the Massachusetts Institute of Technology in Boston reported evidence of a ten percent decline in the amounts of the chemical in the northern hemisphere since 1980.

singapore in the smog

Tropical smogs may use up hydroxyl, triggering a global buildup of pollution, including greenhouse gases.

future forecast

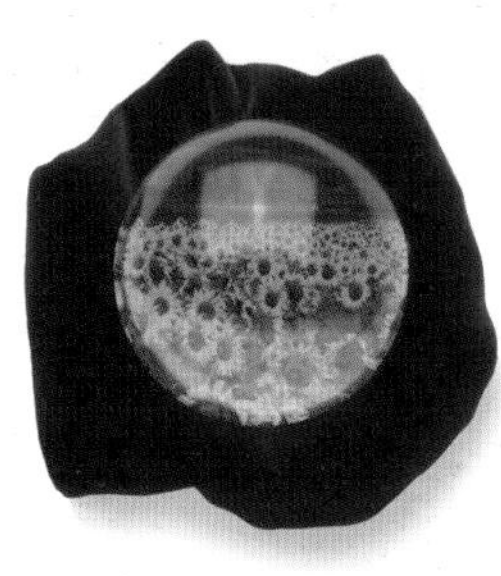

Nobody can predict the future precisely. The latest projections from the world's climate scientists at the IPCC suggest that the average global temperature in the 21st century could rise by between 2.5 and 10.5°F (1.4 and 5.8°C).

The higher estimate represents a change in temperature greater than that between today's temperature and the depths of the most recent ice age (see p.9). But the lower estimate is far less threatening.

how the sceptics see it

Greenhouse skeptics favor the lower estimate, suggesting that the negative feedbacks could be more important than previously realized, and that the warming we have seen to date could be caused, at least in part, by natural factors.

They also point out that while ground-level temperatures appear to have risen in recent decades, the warming has failed to penetrate as far upward into the atmosphere as climate models predict. In fact, at 2 miles (3 km) up, quite large areas of the atmosphere have cooled. Skeptics argue that warming at the surface may be drying out parts of the upper air, reducing the water-vapor feedback and causing the air to cool. If climatologists are wrong about the water-vapor feedback, their projections for our climate may be wrong.

"The frequency and magnitude of many extreme climatic events increase even with a small temperature increase."

Interngovernmental Panel on Climate Change, 2001

what next?

The factors influencing global warming are complex and interlinked, and their combined effects are hard to predict. The effects of global warming are likely to be highly variable.

recap

Negative feedbacks, which have a cooling effect on the planet, include the absorption of heat from the atmosphere by the oceans, and the reflecting of some heat away from the Earth's surface by particles of pollution in the atmosphere.

weather warning

The intensification of the greenhouse effect threatens to make the world an increasingly uncomfortable place. The detail of how it will influence our climate remains unclear – not least because unpredictable positive feedbacks caused by melting ice, increasing water vapor, and changes to ocean temperatures could all accelerate the pace of warming. But a warmer atmosphere will undoubtedly be a more dynamic atmosphere, given to greater extremes of storm and drought, wind and rain. Wet areas will become wetter; dry areas drier. El Niño and the Asian monsoon will both probably become more extreme, but more unpredictable. Areas now affected by famine will grow less food, while many rich lands will grow more. A few ecosystems will simply migrate with the climate; many will die where they stand. Coral reefs and mangrove swamps, tropical rain forests and alpine fastnesses will all disappear. Extinctions will proliferate. And the world will face a flood of refugees from climate change, rising sea levels, epidemics of disease, and water wars.

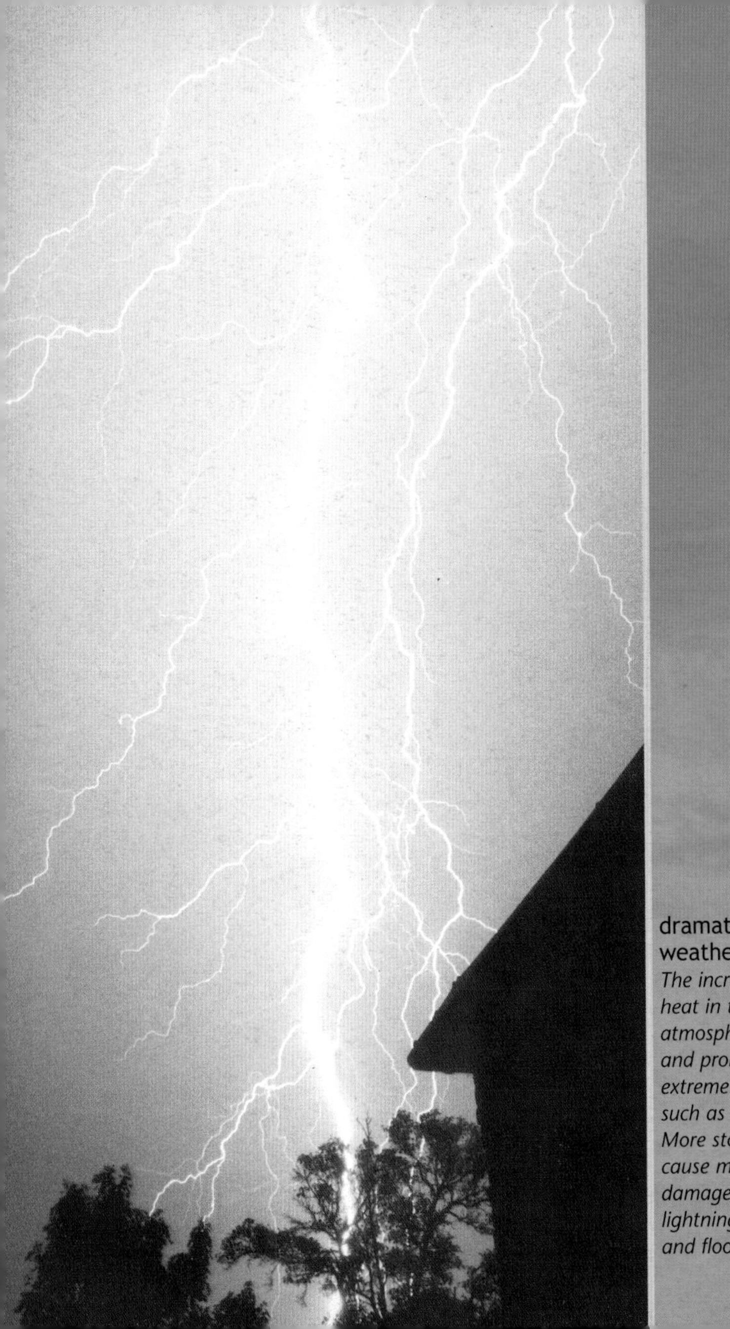

dramatic weather

The increase in heat in the atmosphere will fuel and prolong extreme weather, such as storms. More storms will cause more damage from lightning, wind, and flooding.

climate change

Climate change will prove very variable. Most places will become warmer, but a few will become cooler. Some of the fastest warming is likely to occur in Canada, Russia, and Scandinavia, partly because the positive feedback caused by ice melting (see p.21) will be most intense in those places. That sounds like good news: crops and trees will grow better, heating bills will go down. But the bad news is that nearly all land areas, from the hottest to the coldest, are expected to warm more than the global average. The warming will be most intense in the continental interiors because the circulation of the oceans will act as a moderating influence in coastal areas.

keeping cool
The oceans will draw heat away from the surface in coastal areas – or, at any rate, those that are left after sea levels have risen.

the hot get hotter

Some of the hottest regions today face some of the biggest temperature increases. A large area of Asia, from western China to Saudi Arabia, which regularly sees temperatures above 104°F (40°C), faces further rises of 12.6°F (7°C) by 2100. North Africa and southern Europe are also likely to see major warming. Countries with a strong marine influence and equable climates today – such as Ireland, New Zealand, and Chile – are in line for much smaller changes.

Other general global trends, many already evident now, include greater warming at night and during the winter. This suggests less winter snow and more rain, plus longer frost-free growing seasons in many middle latitudes.

the collapse of the gulf stream?

The Gulf Stream is part of an ocean circulation system in the North Atlantic that is currently driven by the formation of ice in the Arctic. It swathes Western Europe in warm waters, especially in winter, and keeps temperatures higher than they are elsewhere on the same latitude. Scientists at the Potsdam Institute for Climate Impact Research in Germany predict the possible collapse of the Gulf Stream due to global warming. Much of Europe would cool sharply as a result.

current gulf stream mechanism

This ocean cross-section shows how, when ice freezes, it leaves dense saline water that sinks to the ocean floor ("deep water formation"), letting warmer water flow in from the tropics.

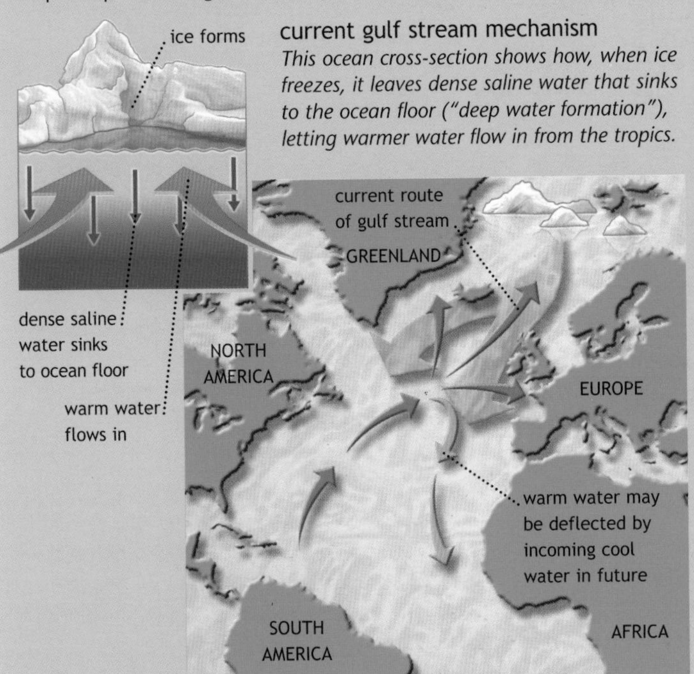

possible changes to the route

Scientists predict that, as the world warms, less ice will form. This, combined with more fresh water flowing into the Arctic from nearby land areas, could shut down the "deep water formation" mechanism that creates the Gulf Stream. In early 2001, Norwegian studies provided evidence that north-flowing currents in the region have already diminished by 20 percent since 1950.

upset hydrology

It will not just be temperatures that change over the next century. In many places the most obvious differences will be alterations to the hydrological cycle – the circulation of water among sea, atmosphere, and the Earth's surface – and hence to patterns of rainfall, floods, and drought, to river flows and vegetation. Water will disappear from where it is expected and needed, and reappear where it is unexpected or only causes chaos. As warming makes the atmosphere more energetic, rates of evaporation and cloud and storm formation will all increase, although the overall effect of these changes will vary greatly according to location.

not a drop
Faltering rainfall is already emptying taps and irrigation canals from North Africa and central Asia to southern Europe.

drying out

Greater evaporation will probably cause the interiors of continents to dry out over the next century. Deserts will spread; oases will die; and river flows will diminish, sometimes with catastrophic results. Nobody can accurately predict future river flows, but at least one study suggests a 40 percent decline in the flow of the Indus River – the sole source of water for Pakistan's and the world's largest irrigation system. The same study predicts a 30 percent loss in flow in the Niger River, which waters five poor and arid countries in West Africa, and a ten percent decline in the Nile, lifeblood of Egypt and Sudan.

spreading sands
The Sahara desert is currently expanding as rainfall diminishes in much of West Africa.

Central Asia can expect a further drastic decrease in the flows of rivers draining into the Aral Sea (see p.30), which is already virtually drying up because of diversions for irrigation. Other inland seas at risk include the Caspian Sea, the Great Salt Lake, and Chad, Tanganyika, and Malawi lakes in Africa.

the greening sahara?

The Sahara has two potentially stable states. Rock paintings there show that in the past it was a cattle-grazing region. And fossilized pollen shows it had woodlands, rivers, and lakes. It turned to desert within a few decades around 5,500 years ago, and could change back just as swiftly, according to some researchers. The region is on a knife-edge because its vegetation is dependent on reinforcing feedbacks between the atmosphere and vegetation. The current state, with little vegetation, produces little rain. Only a small increase in rainfall (caused by global warming), and so vegetation, could turn the Sahara back into woodland.

the sahara today

The Sahara's arid landscape currently contains little moisture; so there is little evaporation and no rain. Most climate models suggest that the Sahara will become even drier, causing desertification in nearby areas.

the sahara of tomorrow?

If the Sahara became covered by vegetation, the landscape would absorb more moisture, resulting in more evaporation and rain-giving clouds.

aral sea drought
The Aral Sea was once the world's fourth-largest inland sea. Irrigation systems upstream have hugely reduced the lake, and its salinity has tripled. Fishing in the sea has ceased. Global warming could make things even worse.

Climate models also suggest the likelihood of more droughts in Europe, North America, and western and central Australia. Some Australian rivers could lose half their flow as the outback becomes more arid.

Currently, 1.7 billion people live in countries that hydrologists describe as "water stressed" because they use more than a fifth of all the water theoretically available to them. It is estimated that this number will rise to 5 billion by 2025. This raises the specter of water wars as countries fight to control this most precious of all natural resources.

increases in flooding

Faster evaporation will also tend to increase the amount of moisture in the air at any one time. The extra heat and moisture will generate more intense tropical storms. There will be more rainfall in some places, particularly in coastal regions and along the lines of storm tracks. Average annual rainfall globally increased by ten percent during the 20th century. Some models predict that extra storms in the floodplain of the Mississippi, for instance, could cause this river to become still more prone to flooding. The Caribbean, Southeast Asia, and other regions already prone to hurricanes and cyclones can expect yet stronger winds, yet heavier rains, and many more flash floods.

Parts of Asia's monsoon rain system may grow more intense. But the monsoon will also grow less predictable and may fail more often. Some climate models suggest that, with more warmth in the tropical atmosphere and ocean, El Niño (see panel, right) will become a near-permanent phenomenon.

disease
A warmer world will allow mosquitoes to bring diseases such as malaria and dengue fever to countries outside the tropics.

what is el niño?

El Niño is a natural phenomenon that has been traced back for thousands of years, a periodic reversal of the winds and ocean currents across the tropical Pacific Ocean, lasting for nine months to a year. This drags rain systems from Asia, causing droughts in normally wet areas, such as Indonesia and parts of Australia. Meanwhile, normally placid South Sea islands see storms, and the Pacific coast of the Americas – usually very dry – suffers storms and floods.

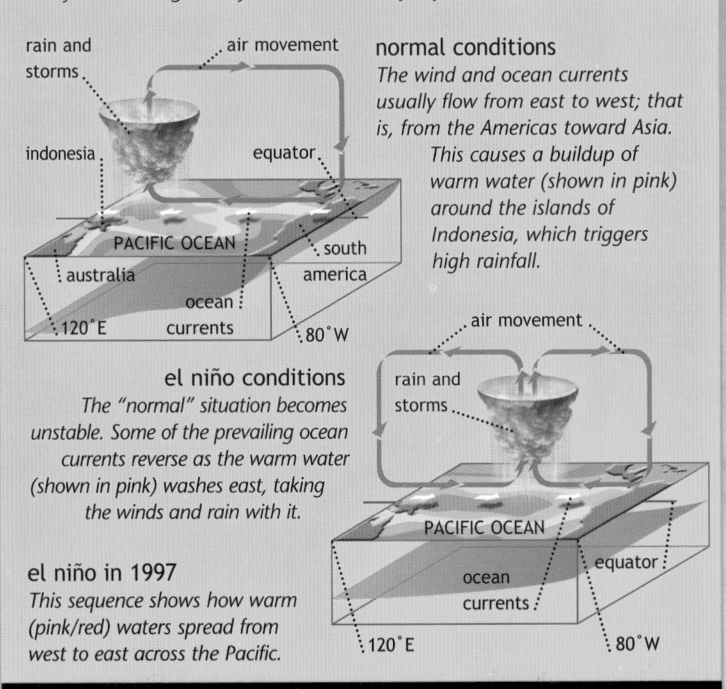

normal conditions

The wind and ocean currents usually flow from east to west; that is, from the Americas toward Asia. This causes a buildup of warm water (shown in pink) around the islands of Indonesia, which triggers high rainfall.

el niño conditions

The "normal" situation becomes unstable. Some of the prevailing ocean currents reverse as the warm water (shown in pink) washes east, taking the winds and rain with it.

el niño in 1997

This sequence shows how warm (pink/red) waters spread from west to east across the Pacific.

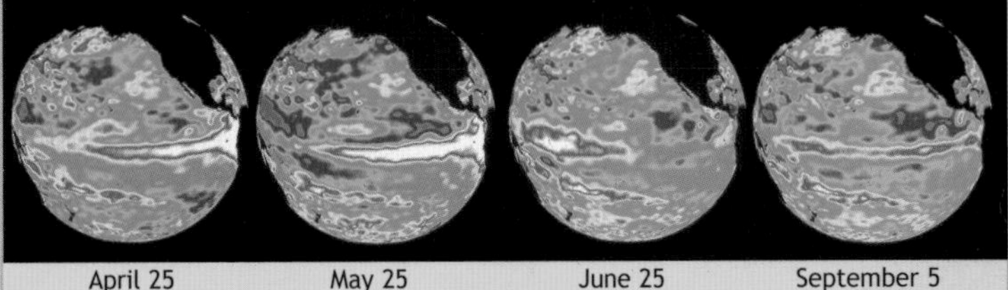

April 25 May 25 June 25 September 5

meltdown

last of the glaciers
Like popsicles in the summer sun, most of the world's glaciers may have melted away by the end of this century.

The melting of glaciers, which is already well under way across much of the planet, seems set to continue, threatening irrigation, navigation, and hydroelectric developments. The Himalaya mountains, for instance, currently have 1,500 glaciers covering 12,750 sq miles (33,000 sq km). Most may be gone by 2100. European scientists say 50–90 percent of Alpine glaciers will also be gone. On current trends, Australia's Snowy Mountains can expect to be snow-free by 2070; and the snows of Mount Kilimanjaro could have all melted by 2015, along with most of the tropical Andean glaciers.

slippery slope
The Athabasca Glacier, part of the Columbia Icefield in Canada, one of many that is melting due to global warming.

changes in precipitation

This global melting will usually be exacerbated by a decline in snowfall and an increase in rainfall. The combined effect is likely to be complex and sometimes sudden. In many parts of the world, rivers freeze up in winter as precipitation accumulates in snow packs in the mountains. In the future, many of these rivers will continue to flow throughout the winter – often at levels higher than any seen before. Meanwhile, spring meltwaters – generally the peak flow of the year – will first surge as the glaciers melt, and then decline dramatically when the glaciers are gone. The consequences downstream are likely to be severe.

In California, the spring meltwaters from the Sierra Nevada snow pack today sustain summer irrigation of crops and lawns in the desert lowlands. But the snow pack is predicted to diminish by 80 percent by the end of the 21st century. As the rushing meltwaters become a trickle, some of the most productive agricultural lands in the world could dry up. Unprecedented floods that washed away communities in the Alps in the winter of 2000 could be a sign of things to come. La Paz, Lima, and Quito – the capitals of Bolivia, Peru, and Ecuador, respectively – depend on Andean glaciers for secure water and hydroelectric power. If the glaciers disappear, as seems likely within little more than a decade, they will need alternative sources of both. In south Asia, half a billion people depend on the glacier-fed flows of the Ganges, Brahmaputra, and Indus rivers.

meltwater flood

In 1994, the unusually full Lake Tsho in northern Bhutan unleashed a flood of glacial meltwater that drowned 27 people downstream.

effects of melting ice caps

As glaciers and the great ice caps of Greenland and Antarctica start to thaw, an increasing amount of extra water will flow into the oceans from melting ice on land. This will contribute to sea-level rise (see p.35).

There is considerable uncertainty about how much melting ice caps will raise sea levels over the coming decades, however. This is partly because nobody quite knows how ice sheets will respond to fast warming at the surface, and partly because warmer temperatures will increase evaporation from the oceans, and may produce more snowfall, particularly in Antarctica, much of which is climatically a desert at present. The extra snowfall

key points

- Melting sea ice, such as that over the Arctic Ocean, does not add to sea-level rise, unlike that from ice caps and glaciers.
- Increased snowfall rates in some places may help compensate for the losses from melting.

would add to the snow pack and might counterbalance losses from melting. Such uncertainties explain why predictions for sea-level rise over the current century vary from 5½in (14cm) – little different from the current rate of rise – to 32in (80cm).

long-term consequences

Once large ice caps start to melt, this process will carry on long after the warming that triggered it has ceased. Even quite modest warming, scientists from Britain's Meteorological Office now believe, will melt most of the world's glaciers and destabilize major ice sheets on land. Just three degrees of warming (likely within a century) "would lead to virtual complete melting of the Greenland ice sheet." It would probably take 1,000 years or more, but this alone would cause sea levels to rise by 20ft (6m).

There are two other major ice sheets on land: the West and East Antarctic ice sheets. The eastern sheet is the largest, but is thought to be stable unless temperatures rise by 27–36°F (15–20°C). But the western sheet, which also has enough water in it to raise sea levels by 20ft (6m), is perched on a large submerged archipelago of islands. Ocean water flows underneath much of it. In the past, scientists have feared it could collapse within a few decades if it were bathed in warmer water. The current view is that this is unlikely unless ocean temperatures rise by more than 18°F (10°C). But not everyone agrees.

arctic meltdown

These massive ice floes off the coast of Spitsbergen, Norway, originated in the Arctic ice sheet. Arctic ice has diminished by 40 percent in 40 years.

rising sea levels

The thermal expansion of the oceans in the 20th century (see p.8) will continue. Warming today will still be causing the tides to rises centuries hence. This is because it takes hundreds of years for warming at the ocean surface to penetrate into the depths. According to a study by Britain's Meteorological Office, even 500 years after temperatures in the atmosphere have been stabilized "sea-level rise from thermal expansion may have reached only half its eventual level." Warming from a doubling of carbon dioxide levels from pre-industrial concentrations would eventually raise sea levels by about 6ft (2m) from thermal expansion alone.

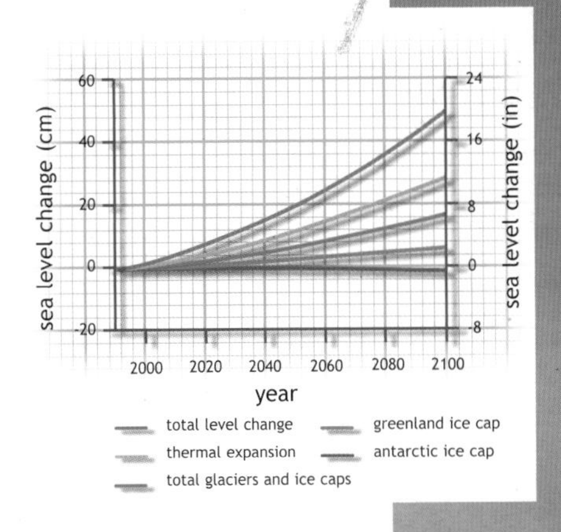

rising sea levels
This graph shows the projected rises in sea levels due to various causes, as well as the total estimated rise. Its projections are an average between the best- and worst-case scenarios currently predicted by the IPCC.

holding back the tide

Sea-level rise is a threat to hundreds of millions of people around the world, from the Nile delta in Egypt to the Baltic Sea shores of Poland and the palm-fringed islands of the South Pacific. An 32-in (80-cm) rise (the highest predicted by the IPCC by 2100) would inundate two thirds of the land area of two far-flung island groups in the Pacific: the Marshall Islands and Kiribati. The Maldives, in the Indian Ocean,

> **"Rising sea levels could annihilate our islands as effectively as an atomic bomb."**
>
> Tom Kijiner, foreign minister of the Marshall Islands, 1994

would not be far behind. Already, high tides sometimes cover the airport runway of the Maldives capital, Male. If researchers are correct in their predictions of centuries of sea-level rise, many more islands are doomed.

Many coral islands also face increased storminess, lost farmland and tourist beaches, and the invasion of salty sea water into their shallow underground water supplies. Their small populations and large coastlines make physical protection using sea walls impracticable. Their only possible salvation could be the ability of the coral itself to continue to grow vigorously, keeping pace with the rising tides. But the vulnerability of coral to bleaching (see panel, right) makes that unlikely.

threats to coastlines

Many coastlines of continents are equally at risk – and contain much larger populations. Coastal regions are already home to half the world's population and, because many of the world's largest urban areas are coastal, they have population growth rates double the global average. The most vulnerable coasts include the lagoons and coastal swamps of western Africa from Senegal to Angola, the northern shore of South

coast at risk
This map of Bangladesh shows how much of its land could be affected by various sea-level rises possible over the next three centuries (oceans are likely to continue to rise long after atmospheric warming has ceased to occur).

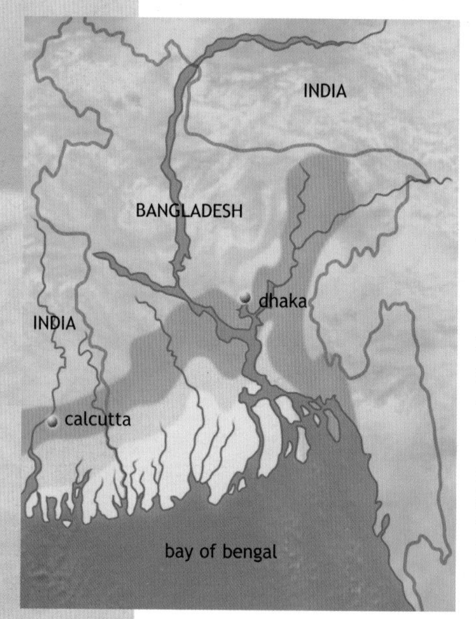

sea-level rise

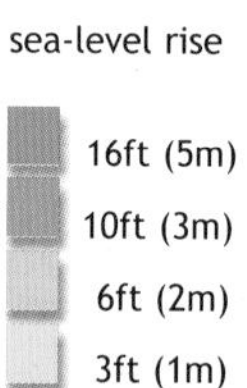

coral bleaching

Coral is not rock, but living matter, consisting of colonies of small invertebrate animals. When it dies, its skeleton forms the reef structure on which new coral grows. When alive, it depends on tiny algae that live inside the coral and provide most of its food and energy – and color. Warm waters kill the algae, causing coral to lose color, known as "bleaching." Occasional, reversible bleaching is a natural feature of many coral reefs. But if the bleaching lasts for longer than a couple of months, the coral starves and dies. As coral dies, the entire reef structure disintegrates. That happened over large areas of the Indian and Pacific Oceans in 1998. As the coral weakens and dies, it is unable to shield the shoreline from rising tides, compounding the effects of sea-level rises.

bleached coral
A sudden rise in sea temperatures kills the algae that give coral its color.

America all the way from Venezuela through the Guyanas to Recife in Brazil, almost the entire eastern seaboard of the US, and much of the coastlines of Indonesia, Pakistan, and the countries around the North Sea.

A recent study found that a 3-ft (1-m) rise would swamp 45,000 sq miles (120,000 sq km) of Asian coastal plains occupied by 44 million people. Some studies suggest that the huge volumes of silt carried by rivers such as the Ganges could raise the delta enough to prevent much sea encroachment. But this is far from clear, and many of these rivers are being dammed, so their silt never reaches the sea.

The Asian study did not include China, whose eastern coastline from Shanghai to Tianjin is extremely vulnerable. Consider the case of Tianjin, a city of 10 million people. Most of the city and surrounding area has sunk by almost 10ft (3m) over the past 40 years as water has been pumped from beneath the city to fill its taps. Parts of the city are currently below sea level. Meanwhile, sea levels are rising by about 1/16in (2mm) a year.

"Some of our islands are only a few meters wide in places. Imagine standing on one of them with waves pounding on one side and the lagoon on the other. It's frightening."

Nakibae Teuatabo, chief climate negotiator, Kiribati, 2000

The eastern US shoreline has 22,500 sq miles (58,000 sq km) vulnerable to a 3ft (1m) rise, 80 percent of it in Florida, Louisiana, Texas, and North Carolina. According to the Federal Emergency Management Agency, a quarter of all the houses on the US coastline could be lost in the next 60 years, with 10,000 homes a year disappearing underwater within a decade.

Some of the most rapid inundations will probably occur on polar coastlines, where the effects of sea-level rise will be accentuated by the thawing of permafrost, which will lower land levels, and the melting of sea ice that currently protects low-lying coasts from wave erosion.

liberty under threat
The eastern coast of the US will be diminished as tides start to rise.

eastern seaboard

sea defenses

Many rich countries believe they will successfully be able to defend themselves against the rising tides – albeit at significant cost. But often their technical skills at doing this in the past mean these countries are already highly vulnerable. Countries such as the Netherlands, Britain, Denmark, Germany, and Poland all protect large areas of coastline that are below current sea levels. The bill for protecting the Netherlands from sea-level rise in the next hundred years has been put at $3.5 trillion. A major Europe-wide study in 2000 predicted that the number of people at risk of flooding in northern Europe would increase tenfold by 2080.

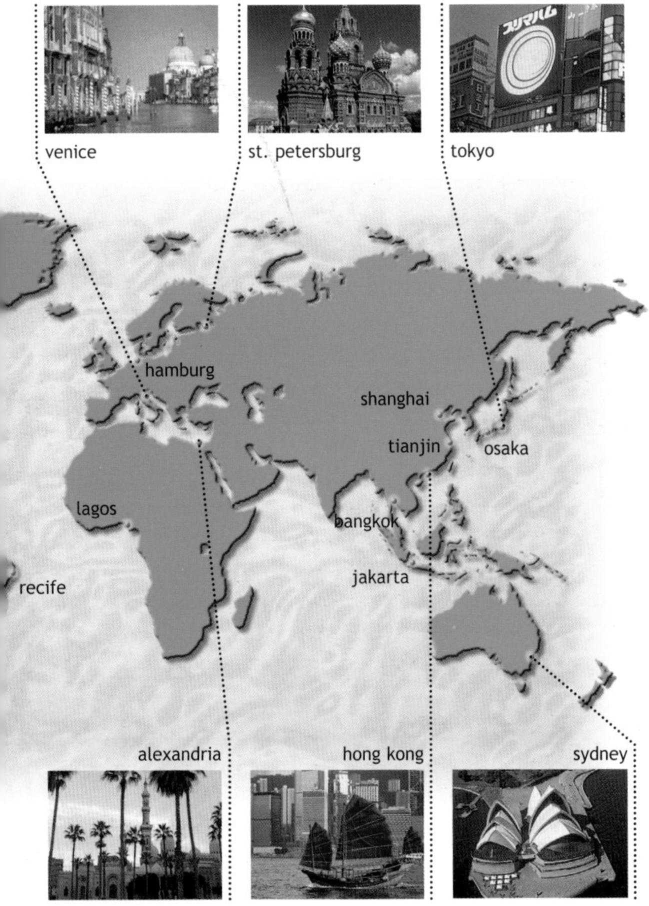

venice

st. petersburg

tokyo

hamburg

shanghai

tianjin

osaka

lagos

bangkok

jakarta

recife

alexandria

hong kong

sydney

recap

Sea-level rise is caused by thermal expansion of the oceans – any liquid takes up more space if it is warmed. And melting ice on land adds to the amount of water in the oceans. Rising tides will inundate low-lying land, speed up coastal erosion, engulf coastal defenses, and pollute underground water reserves with salt water.

ocean invasion

Many of the world's largest and most famous cities face flooding as sea levels rise. Saving them could cost trillions of dollars.

storms

barrier breached?
London's Thames Barrier is designed to protect against 20th-century tides and storm surges. Soon it will not be enough to counter higher tides and worsening storm surges.

The twin perils of global warming – a more volatile climate and rising sea levels – will come together most potently in coastal regions, where rising waters will be compounded by the growing intensity of storm surges. Cities such as London, Hamburg, and St. Petersburg sit on estuaries exposed to storm surges. These are super-high tides that can occur when air pressure is low (which raises sea levels slightly) and offshore winds push the seas into enclosed bays and estuaries. Such surges are predicted to become more frequent and more severe. The IPCC calculates that by 2080, the number of people whose houses are flooded each year could rise from a few million today to up to 200 million.

Much of the coastal world faces an increased risk of severe storms. The IPCC concluded in 2001 that "the most widespread risk to human settlements from climate change is flooding and landslides, driven by increases in rainfall intensity and, in coastal areas, sea-level rise."

extreme weather systems

Tropical cyclones will become more intense, because the heat that gives them energy will be available in greater supply. The result will be stronger rains, higher winds, more vicious storm surges, and greater disruption inland. Watch out for more hurricanes – and huge destruction as floodwaters rush through poorly drained areas of housing and landslides devastate deforested hillsides.

In October 1998, Hurricane Mitch caused floods and landslides throughout Honduras, leaving more than 10,000 people dead and two million homeless. It dumped a year's rain on the small Central American country in a few hours. Many meteorologists believe that Mitch was a product of global warming. Others argue that, whatever its specific cause, Mitch was a sign of things to come for the hundreds of millions of inhabitants of flood-prone river valleys and coastal plains across the world, for those living on deforested hillsides prone to landslips, and for many more.

Storms will wreck many mangrove swamps that protect low-lying coastlines in Asia and Africa from the worst weather coming off the sea. Most of the Sundarbans – the world's largest region of mangrove swamps, covering 2,300 sq miles (6,000 sq km) on the seaward edge of the Ganges river delta in Bangladesh – would be killed by a 18-in (45-cm) rise in tides, taking rare species such as the Bengal tiger with them.

economic losses
The rising cost of weather-related natural disasters worldwide is clear in this graph. Total costs globally were five times higher in the 1990s than in the 1980s.

hurricane mitch
Hurricane Mitch brought record rains and high winds through Honduras in 1998. Some blame global warming.

ecosystems

Ecologists fear that the impact of global warming on biodiversity could be one of the most destructive processes seen on Earth since the start of evolution. As always, there will be winners and losers among individual species. But the speed of climate change is likely to outpace nature's ability to adapt. Ecosystems, and carefully established crops, finely tuned over time to their environments, could be destabilized. India's rice crop – one of the great success stories of the green revolution of the past 50 years – faces heat stress, water shortages, and worsening crop diseases.

rain forests

requiem for the rain forests
Many forests are doomed as global warming dries out continental interiors.

In the Amazon rain forest, the home of more species of plants and animals than anywhere else on Earth, predictions are for higher temperatures and a longer dry season each year. These can be expected to cause widespread damage to the forest – not least since the loss of forest will itself tend to reduce the amount of rainfall in the area. This is because, in a region such as the Amazon, evaporation of water from the canopies of rain forests, particularly in coastal regions, is a major source of rainfall for the forests inland. The rain and the forests are interdependent: if the coastal forests are lost, the interior will start to dry out.

The dangers of a breakdown of rain forest hydrological and ecological systems have been made greater because human activity is already fragmenting the forest at an accelerating rate, making it harder for species under threat to migrate, reducing rainfall levels, and increasing the number of fires that start. Fires could ultimately become catalysts for the forest's destruction. The result of that destruction is likely to be the emergence of scrub or even desert in place of rain forest. According to climate models developed by British scientists, most of the Amazon rain forest will disappear around the middle of the 21st century as a result of such changes. The Amazon will not be alone. Other tropical rain forests face similar devastation.

deserts in the amazon

Studies at Britain's Meteorological Office suggest that, as Brazil gets hotter and drier, it could reach a threshold beyond which the world's largest rain forest succumbs to massive fires from which it cannot recover. Deprived of vegetation, the region will dry further and the thin soils will erode away. The result, by the end of the century, could be close to desert.

loss of diversity

Habitat fragmentation in the Amazon rain forests already threatens many species. A total change of habitat, such as desertification, would destroy many more.

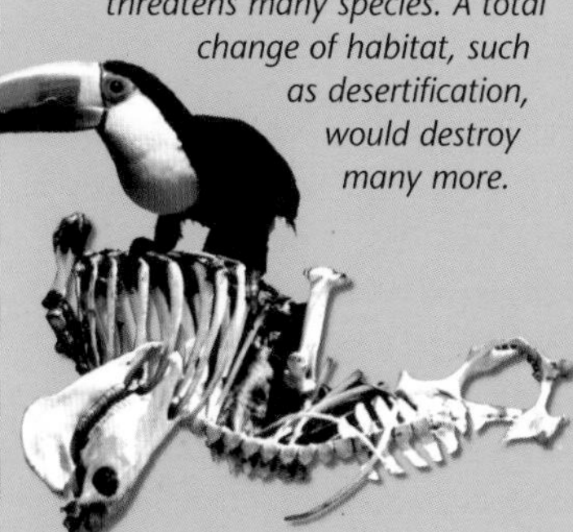

rain forest loss

Human activity has already reduced the Amazon rainforest. Deforestation is adding to the carbon dioxide in the atmosphere. If the whole forest went, global temperatures would rise by 3.6°F (2°C).

migrate or die?

In theory, species and ecosystems can migrate along with the changing climate, but in practice many ecosystems will find it hard to move in step. Birds and flying insects might be able to shift swiftly, but the plants on which they feed may not. Animals and plants do not just need particular climatic conditions to survive; they also need soils, space away from human developments, and corridors of unspoiled terrain through which to migrate. In our crowded world, where natural habitat is fragmented by cities, roads, and farms, it is more likely that some aggressive and adaptable species will invade new territories – creating new mongrel ecosystems of degraded scrubland – while other species will simply die out. Threatened species will probably include the gorillas in the mountains of central Africa, the spectacled bear of the Andes, and the Bengal tiger, already under siege in the mangrove swamps of the Sundarbans as the Bay of Bengal invades.

mountain gorillas
As the mountains lose their forests, mountain gorillas will have nowhere to go.

bengal tiger
Rising sea levels could drown the last home of the Bengal tiger.

fire hazard

As forests are turned into tinderboxes, grasslands will gain the upper hand. And fire will increase its influence elsewhere. In many regions the potential "fire season" is likely to extend by about a month. The arid woodlands around the Mediterranean and the forests of the Himalayan mountains, home to a tenth of the world's mountain plants and animals, are both expected to burn, as are the endless birch and pine forests of Siberia, where fires can already last for weeks without anyone noticing. Here, higher temperatures will trigger more storms, and lightning strikes will do the rest. A Canadian government study forecasts a 40 percent increase in lightning strikes in that country's huge northern forests. The fires will release more greenhouse gases into the atmosphere, accelerating global warming.

forests burn
Fires triggered by lightning are likely to increase. In a warmer climate, the forest is unlikely to regenerate, but may be replaced by grass or scrub.

Away from such destructive forces, warmth-loving species will often gain terrain, while cold-lovers lose out – especially where they have few chances to move. But warmth-loving species may still suffer if they are not tolerant of droughts or, in coastal areas, of flooding.

wetland habitats

Wetlands are another obvious casualty of climate change. Bogs, marshes, coastal lagoons, and many similar features rely on a complex interplay between land and water. Even small changes in hydrology could prove fatal. Some wetlands will be flooded; others will dry out. Clearly, new wetlands will form eventually, but they take time to develop, and the overall extent of replacement wetlands seems uncertain while fast climate change continues. Meanwhile, wildlife will suffer. In Asia, sea-level rise will rob migrating birds of vital refueling stops on coastal marshes in the Mekong and Yangtze deltas. Inland, wetlands may disappear as river flows diminish or change their flooding patterns.

marsh flood
India's Rann of Kutch coastal marsh, home to one of Asia's largest flamingo colonies, is threatened by rising sea waters.

nothing is inevitable

Halting global warming will be hard. It will require the eventual cutting of pollution from burning fossil fuels to a fraction of current levels. A fair world will require the biggest cuts from the nations with the highest pollution levels today, particularly America, Australasia, and Europe. But there is hope. We are using energy ever more efficiently. Greenhouse-friendly energy technologies, such as wind and solar power, are close to being competitive with even the cheapest fossil fuels. Carbon trading under the Kyoto Protocol could provide strong economic incentives for clean energy. Soon, most electricity could come from solar energy, and most cars could be powered by hydrogen, while newly planted forests and even specially fertilized oceans could help soak up the carbon dioxide put into the air by previous generations. The stakes could not be higher. We have the technology to head off perhaps the worst threat to human civilization since the end of the last ice age. But do we have the political will?

wind farm

Clean-energy technologies, such as wind power, are close to becoming cost-effective alternatives to traditional energy generation. They could save the world from global warming.

setting targets

The first international effort to tackle climate change came with the signing of the UN Framework Convention on Climate Change at the Earth Summit in Rio de Janeiro in 1992. Industrialized nations agreed to stabilize their emissions at 1990 levels by the year 2000. Many failed.

By 2000, US emissions were 13 percent higher. The European Union (EU) had kept marginally below 1990 levels, with large reductions in coal burning in Britain and Germany counterbalanced by big increases in countries such as Spain and Ireland. The period coincided with the collapse of industry in the former Soviet Union and Eastern Europe, reducing emissions in many of these countries by 30 percent. Overall, global emissions rose by six percent during the 1990s. The US alone was responsible for half the global increase during the 1990s, exceeding the combined increases of China, India, Africa, and Latin America.

kyoto

The next effort to agree cuts to emissions was the Kyoto Protocol, signed by most nations in 1997. It agreed cuts in emissions of six main greenhouse gases by most industrialized nations, by 2010. Targets averaged a cut of just over five percent cut from 1990 levels, but ranged from an average eight percent cut for the EU to an eight percent increase for Australia and a ten percent increase for Iceland, both of which successfully argued that they had special needs. The US promised a seven

a fatal dependency
The burning of coal, oil, and gas is the major culprit in the rise in emissions of carbon dioxide into the atmosphere.

the cost of kyoto

Will meeting the Kyoto emissions targets cripple economies? The European Union (EU) thinks not. It reported in 2001 that it could meet its Kyoto commitments twice over at minimal cost using a combination of cost-effective new green technologies, ranging from biofuels to encouraging energy-efficient buildings, and by providing incentives through carbon trading. Meeting the target itself would cost just 0.06 percent of GDP, almost half previous estimates.

percent cut. Russia was allowed a "no-change" target, even though its emissions had plummeted since 1990. There were no targets for developing countries, whose per-capita emissions are mostly well below those of the industrialized nations.

At the Kyoto meeting, it was also agreed to develop a series of "flexible mechanisms" designed to make it easier and cheaper for countries to meet the targets. These included provisions for countries to plant and manage forests to soak up carbon dioxide and so offset increased emissions (see p.61), to trade in rights to pollute, and to gain credit for investing in cheap emissions-reduction projects in developing countries. After the withdrawal of the US from the negotiating process in early 2001, a "rule book" for these mechanisms was agreed at Bonn in July that year. Governments declared

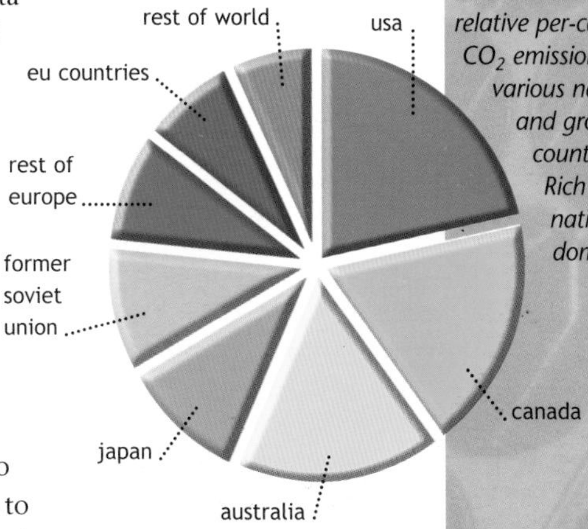

rest of world
usa
eu countries
rest of europe
former soviet union
japan
australia
canada

who pollutes?
This chart shows the relative per-capita CO_2 emissions of various nations and groups of countries. Rich nations dominate.

achieving targets
This graph shows the emissions cuts that would need to be made over the next 50 years to stabilize CO_2 concentrations at various levels. IS92A is a term for the scenario if emissions continue at the current rate.

their hope that sufficient legislatures would ratify the deal for the Protocol to enter into legal force by the time of the Earth Summit in South Africa in late 2002.

the longer run

Scientists and policymakers agree that the Kyoto Protocol is just a small step toward the deep cuts in emissions that will be needed in the long run to stabilize greenhouse gas concentrations in the air. Even President George W. Bush agreed that the world should be seeking to set such a limit. But where should that limit be set? One suggested ceiling was double pre-industrial levels of the main gases – this would be about 550 parts per million (ppm). Other suggested targets included the tougher 450 ppm and the more lax 750 or 1,000 ppm.

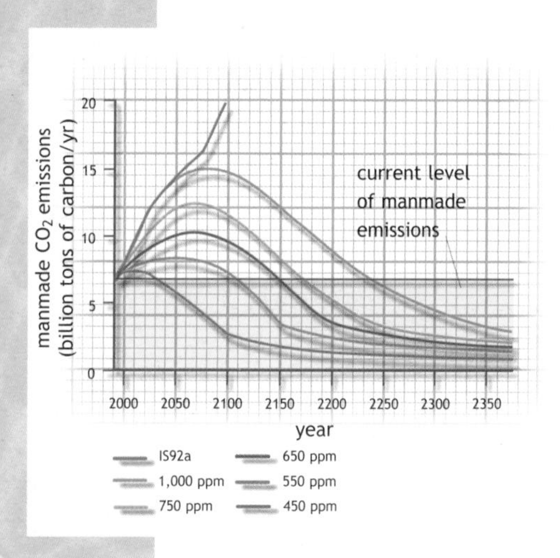

Are the targets achievable? There is time. Studies by the IPCC show that to meet the 450 ppm limit would require global emissions to drop below 1990 levels within 50 years, whereas 750 ppm would give us a century to get below 1990 levels, and a 1,000 ppm target might give us as much as two centuries. But each scenario would require a continued decline in emissions thereafter to the level of natural absorption.

key points

- Today's level of carbon dioxide in the atmosphere is about 370 ppm.
- Levels before the beginning of industrialization were 270 ppm.

how much fuel is left?

In the 150 years since the start of the industrial revolution, the world has released into the air around 290 billion tons of carbon from burning fossil fuels – including some 217 billion tons since 1950. Altogether there is at least 14 times more carbon – 4,000 billion tons – still beneath the ground.

However, there are believed to be only around 300 billion tons of conventional gas and oil in the ground. So the real threat to our future, according to studies by some climate analysts, is from the remaining coal reserves – which probably exceed 3,000 billion tons – and novel forms of fossil fuel, such as oil shales and tar sands, which could add a further 700 billion tons. As the oil runs out, governments will face a stark choice between investing in these "high-carbon" energy sources and taking the "low-carbon" road with renewable sources such as wind, solar, and hydrogen.

building on progress

Despite the continuing industrialization of much of the planet, carbon dioxide emissions are not soaring as fast as you might expect. As we have seen, global carbon emissions increased by just six percent in the 1990s. This compared with a 15 percent gain in the 1980s, 29 percent in the 1970s, and 58 percent in the 1960s.

The recent modest increase in emissions can be put down in part to the collapse of the Soviet and Eastern European economies, but it also reflects two long-term trends: the increasing efficiency with which energy is used in many countries, and a fall in the use of the dirtiest fuel, coal. Globally, the amount of carbon emitted into the air for every dollar of economic benefit gained has fallen by 41 percent since 1950.

avoiding waste
In the West, and particularly in the US, there is still an upward trend in the number of cars per household. This wasteful trend needs to be reversed to help cut emissions.

nuclear option *Countries such as France and Britain would have higher CO_2 emissions without their nuclear power stations.*

increasing efficiency

Along the path to Western-style development, most countries begin by using fuel very inefficiently. Initially, a surge in fuel use causes overall emissions to soar. Yet beyond a certain point, smogs, shortages of resources, and improved technology provide incentives for better fuel efficiency and a move away from coal. Most large emitters are now in this phase.

The latest and most dramatic recent example of this trend is China. It cut its carbon emissions by 18 percent in the late 1990s, at a time when its economy grew by more than 30 percent, largely by shutting coal mines and small, inefficient factories. Its coal use fell by 27 percent from 1996 to 2000. This switch was big enough to cause a small drop in global fossil fuel use in 1999 and 2000. This case is evidence of another hopeful trend. As succeeding countries industrialize, they reach the moment where profligacy turns to efficiency earlier, due partly to greater concern about environmental issues, and partly to the greater availability of cheap, clean-energy technologies.

Optimists believe that, as some fossil fuels become more scarce and expensive, as energy is used more efficiently, and "greener" technologies become more popular, poor countries will industrialize using 21st-century energy technologies without starting on the dirty path of 19th-century Europe.

Nuclear power can be seen as part of the push to efficiency, since it generates electricity without burning fossil fuels. But many environmentalists oppose a nuclear solution to global warming because of fears about nuclear safety, continuing problems with handling and disposing of nuclear waste, and the risk of nuclear proliferation – turning nuclear materials from power plants into nuclear weapons.

eco-friendly energy sources

Wind and solar power are the two mainstream eco-friendly energy options available today. Both are as likely to benefit the developing world as the developed. The blades of wind turbines turn on the plains of India and the steppes of Mongolia, as well as off the shores of Denmark. Solar panels glint on roofs in rural Kenya even more than in sun-drenched Australia. But there are other options to turn natural, nonpolluting substances into energy.

wind power

The worldwide power-generating capacity of wind turbines grew by 21 percent a year throughout the 1990s. By late 2001, Denmark had 5,600 wind turbines providing one tenth of its electricity – a proportion it hopes to raise to 50 percent by 2030. The world's largest offshore wind farm is at the entrance to Copenhagen harbor; its companies have cornered much of the world market in wind turbines.

Meanwhile, Germany, already the world's largest generator of wind power with 9,000 turbines, announced plans to be producing one third of Europe's wind power

wind worries
Wind turbines are criticized for taking up too much land. Their future could lie in offshore wind farms, where there is space to spare and the winds tend to be stronger.

sea breezes
Winds over the sea blow up to 40 percent faster than those over land.

within a decade. It, too, will concentrate on offshore turbines. Spain is on the verge of becoming the world's second-largest generator of wind power. And the British government announced plans for 18 giant offshore wind farms around its coast, as part of a scheme to generate one tenth of its power from renewable sources by 2010.

Wind power is now a practical economic proposition for mainstream power generation. Technical advances have meant that, in the past two decades, the cost of generating electricity from wind has fallen from 40 cents a kilowatt hour to 5 cents, according to the American Wind Energy Association. This makes it close to competitive even with the cheapest commercial energy source – natural gas – which typically costs 3.5 cents. As a result, wind power is taking off in Australia, Morocco, China, Japan, and the US, where farmers in Iowa can make 20 times more by leasing their land for wind turbines than they can by growing corn. The Worldwatch Institute in Washington, DC, calculated that three Midwest states – North Dakota, Kansas, and Texas – could generate enough wind power to meet the electricity needs of the entire US.

"The potential for rapid technological innovation leading to clean energy is clearly extraordinary. Governments need to unleash this potential."

Klaus Toepfer, director, UN Environment Program, 2001

solar power

Solar power grew even faster than wind power – at 30 percent a year – during the 1990s. Photovoltaics, or solar panels, use light-sensitive semiconductors to generate electricity. They can provide power for individual buildings or be assembled in large arrays to supply electricity into a grid system. One British company will supply a typical house with enough roof panels to power the home, with excess to sell to the grid on sunny days. Future buildings will have walls as well as roofs made of solar panels.

the switch from oil

Big oil corporations are becoming providers of solar power, partly as a hedge against future changes, and partly because some genuinely see it as the energy source of the future. Shell Renewables, the green arm of Europe's largest oil combine, plans to spend a billion dollars on solar and wind power between 2001 and 2006. The company already boasts one of the world's largest solar-power development projects – to power poor rural communities in South Africa.

But it has competition from its arch-rival BP (which likes to say that its initials now stand for Beyond Petroleum, rather than British Petroleum). BP has a quarter of the world's solar-power business, and in 2001 trumped Shell with a huge solar-powered project in the Philippines. In remote villages, far from the national grid, 400,000 people should soon be powering their TVs from solar panels.

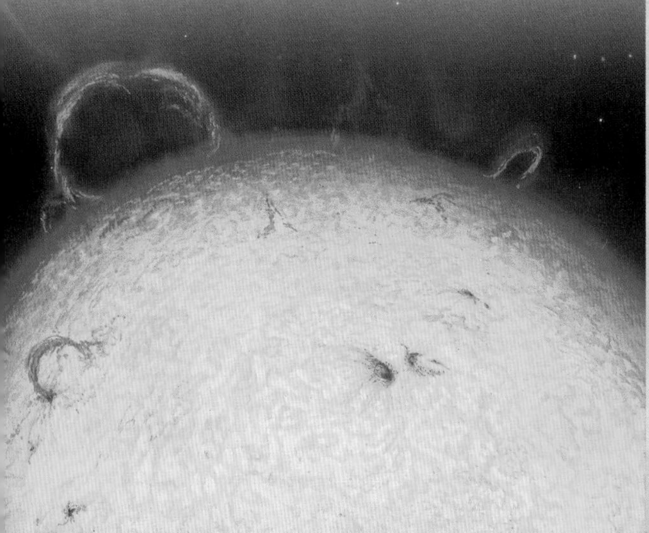

solar power

Energy from the sun can be harnessed in a very direct way by providing a large surface area, in the form of windows or roofing, to collect heat, and insulation to ensure the heat does not escape. It can also be converted, via solar cells, into electricity.

This is undeniably impressive, but both solar and wind power are developing from a very low starting point. Even if their capacity were to increase a further tenfold over the next 20 years, they would still supply only two percent of the power market.

"In the world's most isolated areas, solar is often the most effective way to supply basic, essential electricity needs."

Harry Shimp, BP Solar, 2001

short-term solutions

Climate-friendly biofuels could be a new cash crop by 2010. The thinking is that trees, crops, and farm waste could directly replace coal in power plants or, even better, be distilled to produce ethanol, a more concentrated fuel that can be burned in either power plants or directly in vehicle engines. Provided the fields and forests are replanted each time they are harvested, there is no overall effect on global warming, since the new crop soaks up the same amount of carbon dioxide as is released by burning the harvested crop.

But in the short term, as these technologies develop, the IPCC believes that the world is likely to make do with existing technologies for much of their power. That means using nuclear power (see p.52) and hydroelectric power, despite their environmental drawbacks. Hydroelectric power is produced by damming rivers to store water to be discharged through turbines when power is needed. However, most of the best sites for dams are already in use, and reservoirs take up a lot of fertile land and displace people. Silt is captured behind the dams, making the reservoir useless over time and reducing silt flows downstream, where the silt is needed to bolster deltas and coastlines. In this way, reservoirs add to the threat of sea-level rises. Finally, rotting vegetation in many reservoirs releases significant amounts of greenhouse gases such as methane (see pp.17–18).

king coal

There is more coal in the ground than any other fuel, but to burn it in conventional ways would be disastrous for the climate.

Coal could be replaced with natural gas – efficient gas turbines produce only around half as much carbon dioxide for every unit of power as a conventional coal power station. A combination of these short-term solutions could provide a breathing space for the development of newer technologies.

electricity in the future

In ten years, we could all be making and selling our own electricity. The future of electricity generation could lie in millions of backyard power plants tapping into wind turbines and solar panels, hydrogen fuel cells (see p.60), and even natural gas. This would bring less pollution, greater efficiciency, and an end to power cuts. The developing world probably has the most to gain from this "micro-generation."

home generation

Every new home could soon include a micro-generator, able to generate electricity at prices comparable to the large power stations of the industrial age. Where a national grid exists, the home could export excess energy to the grid on a sunny day and import it when dull.

local power

In developing countries, local networks running on solar cells could provide all the necessary electricity without making use of a national grid. This will make it possible to electrify the developing world while stabilizing greenhouse gases.

transport

Transport is the fastest-rising sector of carbon dioxide emissions. Emissions from vehicles of all sorts are rising annually by 2.5 percent globally, and by 7 percent in Asia, where car ownership is growing fastest. Meanwhile, the US has 4 percent of the world's population, but consumes 43 percent of the world's gasoline.

Real technological improvements in vehicles are being outpaced by soaring vehicle use. Moreover, as the IPCC reports, "improvements in design have largely been used to enhance performance rather than to improve fuel economy." And when vehicles do run more efficiently, they often get snarled up in traffic and idle away the emissions gains because there are too many vehicles on the road. Most difficult of all, we have designed our urban environments around the limitless use of vehicles.

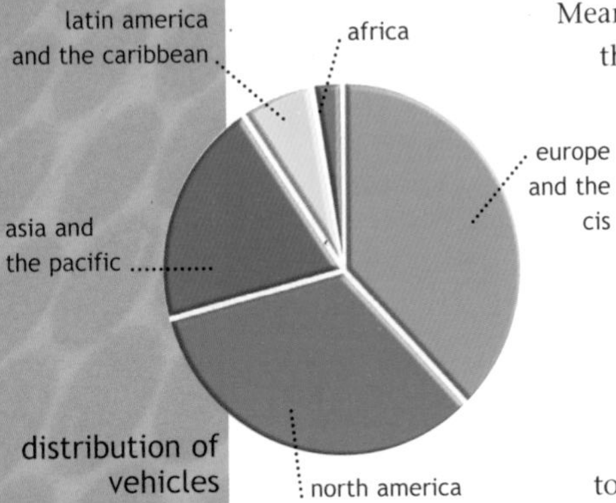

latin america and the caribbean

africa

europe and the cis

asia and the pacific

north america

distribution of vehicles

This chart shows where the world's 680 million vehicles are situated.

a new approach

To make progress in reducing vehicle emissions, we need to start redesigning our cities. Architects such as Britain's Richard Rogers have pioneered schemes for much denser urban development, centered on nodes for efficient urban public transit systems. Even in present-day cities, authorities are investing more heavily in subways, trams, and so on. In the not-too-distant future, long-distance mass movement could be by magnetically levitated

maglev train
A maglev train is suspended and propelled by magnetic fields generated by a rail on which it runs.

("maglev") trains traveling underground in low-pressure tubes, and consuming one tenth as much energy as today's trains. With populations based around a maglev station, the numbers of roads and cars could be reduced in favor of pedestrianized areas.

An alternative future sees us sticking with the car – but in new forms that are virtually pollution-free. To do this, the car would have to be totally redesigned. The Rocky Mountain Institute in Colorado – a think-tank for technologists – has developed designs for an ultra-efficient "hypercar," made of light materials with vastly improved traction and driven by an electric motor for improved efficiency. It would be fueled by hydrogen.

the hydrogen economy

Hydrogen is the ultimate inexhaustible fuel. It can be made from water and burned just like gasoline. But rather than burning it, the key to a future hydrogen economy is likely to be the hydrogen fuel cell – a portable, versatile energy storage medium that produces no pollution (see p.60). The critical question is how the hydrogen is produced. Large amounts of power are

future car
Cars driven by hydrogen fuel cells will be in showrooms by 2004, and many manufacturers believe they will dominate the market within two decades. This methanol-powered fuel-cell car was unveiled in Berlin in 2000.

the hydrogen fuel cell

A hydrogen fuel cell is in some ways like an ordinary battery, with two electrodes (one positive, one negative) in an electrolyte (a solution that conducts electricity). Hydrogen is fed into one side of the cell (the anode) and oxygen into the other (the cathode). A chemical reaction occurs, splitting each hydrogen atom in two. The negatively charged electron is routed through a circuit that, for example, powers a lightbulb. The remaining proton passes through a membrane to the cathode where it rejoins the electron and combines with the oxygen to form water.

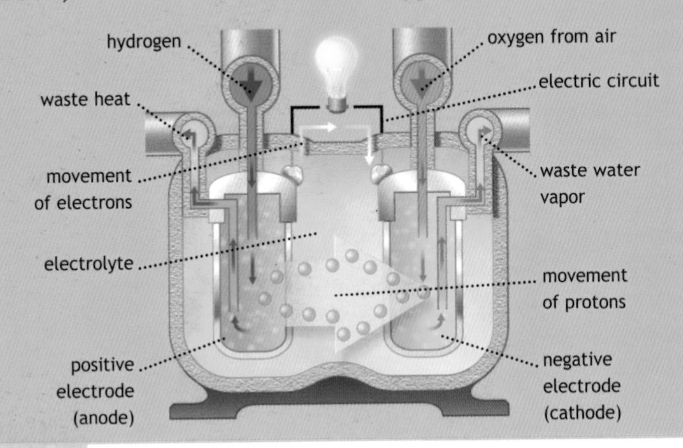

required. If that comes from fossil fuels, the gains are minimal. So a renewable source of energy must be found. Enthusiasts see massive production plants for hydrogen being set up in parts of the world where there is ample water and plenty of large-scale renewable energy available.

Even oil companies and car manufacturers agree this is the way in which the world is heading. But the distribution infrastructure required to create a "hydrogen economy" would be considerable. And until energy from hydrogen is available everywhere, very few drivers will want to switch to a hydrogen-fueled car.

carbon sinks

One way of buying time for reducing emissions of carbon dioxide would be to try to capture some of the emitted gases through biological carbon "sinks" such as forests, the main natural reservoir for carbon on the Earth's land surface. Forests in Europe and North America already appear to be soaking up significant amounts of carbon dioxide. This is partly because new forests are being planted, partly through faster growth of existing forests in the warmer, carbon dioxide-rich air, and partly because many forests are still regrowing after poor management in the past.

carbon sink
Carbon can be taken out of the atmosphere by planting trees. As they grow, the trees absorb carbon dioxide to make leaves and branches.

pros and cons

The IPCC estimates that sink projects could absorb up to 100 billion tons of carbon in the first half of the 21st century, or between ten and 20 percent of expected emissions from fossil fuel burning during that period.

But it warns that land could be in short supply for large sink projects. Also, accounting for carbon in forests is difficult, especially because most of the carbon is held within the soil rather than the trees. And, sometimes, planting trees can release carbon from soils, especially if wet, peaty soils are drained.

"Land carbon sinks may help to reduce greenhouse gas levels in the atmosphere during the short term, but the amounts of carbon dioxide that can be stored are small compared to emissions from burning fossil fuels."

Britain's Royal Society, 2001

key points

- Carbon sinks such as forests can absorb significant amounts of carbon dioxide from the air.
- The trees need to be healthy and growing well for the maximum benefit – if they rot, burn, or die in other ways, little has been gained.

Moreover, some of these forests could end up adding to global warming, according to studies by Britain's Meteorological Office. The problem is that wherever trees are planted on snow-covered land, they replace a white surface that reflects solar radiation back into space with a dark forest canopy that absorbs the heat. In many parts of Canada and Siberia, this warming effect will be greater than the cooling effect from carbon absorption by growing trees. And forest sinks are vulnerable to the fires that are likely to be an increasing feature of future climates.

an answer in the soil

Another idea is to use farm soils to soak up carbon. Soils are a massive carbon store, holding four times as much as the atmosphere and three times more than the world's trees. But the world's soils have lost about 100 billion tons of carbon over the last 10,000 years as a result of cultivation and deforestation. As soils are plowed, roots and other carbonaceous vegetation in the soils are exposed to the air and oxidized to carbon dioxide. This is good for removing weeds – but at the price of lost carbon. Much of that lost carbon could be recaptured, however, by changes in farming methods. One study estimates that 15 years of emissions from fossil-fuel burning could be soaked up in soils over the next 50 years by improving degraded soils, reducing erosion, and finding ways of farming without plowing.

iron filings

Some scientists believe that extra carbon could be captured in the oceans by mimicking what many marine scientists believe is one of the key processes that drives the world into ice ages. Researchers (see panel, right) argue that the amount of iron blowing from the land onto the oceans has varied widely in the past, with greater amounts reaching the sea during cold periods when the land is dry and dusty. They say these variations could have caused a

the iron filings experiment

An international team of scientists has successfully "fertilized" an area of the ocean south of New Zealand with iron filings. They released eight tons of the metal over an area 5 miles (8 km) across. It produced a sixfold increase in plankton growth that spread out over a wide area and lasted for many weeks. The plankton sucked up carbon dioxide from the water as it grew, and the ocean absorbed more carbon dioxide from the air to make up for the loss. The result was less carbon dioxide in the atmosphere. To be effective, this process would need to be repeated at regular intervals.

20 percent rise and fall in atmospheric carbon dioxide levels – a feedback to the cooling and warming at the beginning and end of ice ages. The scientists involved in the experiments have warned that it would be dangerous to use the mechanism to try to head off global warming because of unknown impacts on ecosystems. But experiments are likely to continue.

"By adding iron compounds to the oceans, a 'technological fix' to remove carbon dioxide from the air might be practicable."

John Gribbin, British scientist, 1987

carbon tombs

There may be an even more direct way of reducing carbon in the atmosphere – by capturing it as it leaves the chimneys of power stations and big industrial plants. A modestly sized power station, generating around 500 megawatts, emits about 500 tons of carbon an hour, or 120 million tons in its 30-year lifetime. Technologies exist to remove the gas from the flue emissions and store it under pressure or in liquid form, although they currently consume about a fifth of the plant's electricity.

A big problem is where to put the waste carbon dioxide once it has been collected. Even when liquefied, it takes up more space than the carbon fuel from which it comes. One option is to discharge it into the oceans through long pipes. The liquefied gas would be heavier than water and would form a layer at the bottom of the sea before slowly dissolving.

Another idea for storing waste carbon is to put it underground – in porous rocks such as chalk and limestone. But the most likely approach now appears to be injecting it into depleted oil reservoirs. Oil companies already squirt water into these voids to push out remaining oil. So why not liquefied carbon dioxide?

North Sea oil and gas fields, for instance, could hold 4,750 million tons of carbon. Existing production platforms could be used to pump the gas back into the depths. A full-scale trial of this idea is underway in the Norwegian sector of the North Sea. Statoil, the Norwegian state oil company, will strip carbon dioxide from gas on a production platform on the Sleipner gas field. The carbon dioxide – a million tons a year – will then be compressed and pumped into pores in the sandstone rocks lying beneath the sea, filling voids that were left empty by past gas abstraction.

north sea rig
Oil rigs helped in the creation of our greenhouse problem by pumping out oil. Now they could help solve it by returning the carbon to beneath the waves.

the choice is ours

The greenhouse problem can reasonably be called the most serious threat facing the future stability and habitability of our planet. Humans are now interfering in planetary processes such as the carbon cycle on a scale that runs the risk of overwhelming those processes.

As a species, we might be said to have had a lucky run. There have been no massive meteorite hits on Earth, or truly shattering geological events to knock us off course. Even the climate, it now appears, has been more stable than in most past eras. We cannot insure ourselves against all eventualities in the cosmos, but we can try not to sow the seeds of our own destruction. And the good news is that our technical inventiveness and ability to analyze our predicament does give us the chance, during the current century, to prevent the possibility – even the probability – of a manmade climatic catastrophe. We have the technology. The best evidence is that adopting that technology in a sensible manner does not need to be economically ruinous. What is less clear is whether we have the social and political tools to do the job.

"The task before us is enormous. If we are to bring greenhouse gas emissions down to a sustainable level, we need to make radical changes in the world economy and in the way we all live."

Kofi Annan, UN Secretary-general, 2000

now is the hour

Can we save our climate, or will we allow it to turn our world into a hot, infertile desert? Time is running out.

key points

- Global warming is happening now, but the worst is yet to come.
- Not all the science is clear, but the majority of researchers agree that we must stabilize the levels of greenhouse gases in our atmosphere.
- The technologies are available and not prohibitively expensive.

glossary

biofuel
A fuel derived from living matter. Sources could include anything from farm waste to timber to ethanol distilled from crops.

carbon cycle
The natural exchange of carbon among the atmosphere, oceans, and the Earth's surface. The element may be dissolved in the oceans, absorbed within living organisms and soils, or floating in the air as carbon dioxide (CO_2).

carbon sink
Anything that absorbs carbon dioxide from the atmosphere. Usually applied to natural features of the planet, such as forests, that could be grown or engineered to absorb more carbon dioxide.

carbon trading
Idea, incorporated into the **Kyoto Protocol**, that would allow countries to trade in rights to emit carbon dioxide into the air. The aim is to make measures to reduce emissions more cost-effective.

climate change convention
Officially the UN Framework Convention on Climate Change, signed in 1992 at the Rio Earth Summit. Its signatories, including virtually all of the world's governments, promised to work to avoid dangerous climate change. One of the first outcomes was the **Kyoto Protocol**.

climate model
Normally computerized simulation of the workings of the atmosphere. Used to predict the effect of future changes, such as an accumulation of **greenhouse gases**. The basic tool behind warnings about climate change.

coral bleaching
Loss of color in coral caused by rising sea temperatures. Can signify death of coral.

earth summit
Meeting of the world's governments at irregular intervals to address environment and development issues. The first was in Stockholm in 1972, the second in Rio de Janeiro in 1992, and the third is scheduled for Johannesburg in 2002.

el niño
Periodic switch in ocean currents and winds in the Pacific Ocean. Lasts up to a year, has a huge impact on climate in the region, and produces ripples right around the world. Its frequency and intensity is believed to be altered by climate change.

feedback
Any byproduct of an event that has a subsequent effect on that event or its causes. A positive feedback amplifies it, while a negative feedback dampens it down. Key climate feedbacks involve water vapor, snow and ice, clouds, and the changing growth rates of forests.

fossil fuel
Fuel made of fossilized carbon, the remains of ancient vegetation. Examples include coal, oil, and natural gas.

fuel cell
Like a battery, but with a continuous supply of fuel. Hydrogen fuel cells are seen by many as the portable energy source of the future.

global warming
Synonym for the greenhouse effect and climate change, although the term is not entirely accurate since some parts of the planet may cool.

greenhouse gas
One of several gases, such as water vapor, carbon dioxide, and methane, that trap heat in the atmosphere, rather like the panes of a greenhouse. Their accumulation is believed to be responsible for **global warming**.

hydrogen economy
Hoped-for world in which hydrogen is the prime energy source.

hydrological cycle
The movement of water among the oceans, atmosphere, and Earth's surface (which includes rivers, lakes, vegetation, and soils).

hypercar
Trademark for a design of hyper-efficient greenhouse-friendly car made of light materials, using improved traction and an electric motor powered by **hydrogen fuel cells**.

ice age
Period of global cooling, probably caused by variations in the planet's orbit (see **Milankovitch wobbles**), that have occurred every 100,000 years for the past several million years. They last tens of thousands of years.

ippc
The Intergovernmental Panel on Climate Change comprises some 3,000 scientists appointed by the UN through national science agencies to study the causes, impacts, and solutions to global warming, as well as to advise the policymakers at the **climate change convention**.

kyoto protocol
The 1987 agreement to cut emissions of **greenhouse gases** by industrialized countries by 2010 and to help poor countries choose a more greenhouse-friendly path to development. One signatory, the US, has subsequently renounced the protocol, but it can still come into force when sufficient government legislatures ratify it.

little ice age
Period from the 15th to the 19th centuries when the world was a little cooler than today. Probably caused by a decline in solar heating from the sun.

micro-generation
Small-scale electricity generation that could make large power stations obsolete. Will enable individuals to generate electricity at home and swap it on the national grid.

milankovitch wobbles
Wobbles in the orbit of the Earth that can influence climate over long time-scales. Discovered by Serbian mathematician Milutin Milankovitch. Believed to be trigger for **ice ages**.

ozone hole
Most extreme sign of a thinning of the ozone layer seen in recent decades. Most ozone is lost for a few months each spring over Antarctica and, to a lesser extent, the Arctic. Not to be confused with the greenhouse effect, although some manmade chemicals influence both, and the two effects tend to reinforce one another. The 1987 Montreal Protocol is an international agreement to heal the ozone layer and seal up the holes.

photovoltaic cell
Light-sensitive semiconductors, like computer chips, that can generate electricity. Assembled into solar panels, they can power homes, stores, factories, or even entire electricity grids.

renewable energy
Any energy source that is not consumed by its use. Includes solar, geothermal, wind, and hydroelectric energy, as well as hydrogen and **biofuels** (provided the crops are replanted).

stratosphere
Layer of the atmosphere starting about 6 miles (10 km) up. Home of the ozone layer.

sunspot
Most visible feature of most active phase of solar cycles. Signify eruptions in the activity of the sun that can cause small increases in the amount of radiation reaching the Earth.

thermal expansion
The warming and resulting increase in size of the oceans. Along with melting land-ice, thermal expansion of the oceans is causing a worldwide rise in sea levels.

troposphere
The lowest layer of the atmosphere. The area within which our weather occurs, and within which **global warming** is taking place.

index

a

b

c

d

e

further reading

Global Warming: the complete briefing; John Houghton; CUP 1997; ISBN 0521629322

Climate Impact and Adaptation Assessment; Martin Parry and Timothy Carter; Earthscan 1998; ISBN 1853832669

The Kyoto Protocol; Michael Grubb and Duncan Brack, eds; The Brookings Institution 1999; ISBN 1853835803

Global Warming: can civilisation survive?; Paul R. Brown; Blandford 1996; ISBN 0713726024

Greenhouse: the 200-year story of global warming; Gale E Christianson; Penguin Books 2000; ISBN 0140292586

Late Victorian Holocausts: El Nino famines and the making of the third world; Mike Davis; Verso 2001; ISBN 1859847390

The Carbon War; Jeremy Leggett; Penguin Books 2000; ISBN 014028494X

Contraction & Convergence: the global solution to climate change; Aubrey Meyer; Green Books 2000; ISBN 1870098943

The Manic Sun; Nigel Calder; Pilkington 1997; ISBN 1899044116

internet resources

http://www.gci.org.uk
Website of the independent Global Commons Institute

http://www.unfccc.int
Website of the UN Framework Convention on Climate Change

http://www.climateark.org
Climate Ark offers news archives and extensive links to other sites

http://www.climnet.org
Climate Network Europe, focusses on EU response to global warming

http://www.usgcrp.gov
US Global Change Research Program, official US federal government website

http://www.cru.uea.ac.uk/tiempo
Tiempo Climate Cyberlibrary, run by the Climatic Research Unit of the University of East Anglia

http://greenpeace.org
Greenpeace website

http://www.panda.org/climate
World Wildlife Fund website

picture credits

The publisher would like to thank the following for their kind permission to reproduce their photographs. KEY: a = above; c = centre; b = below; l = left; r = right; t = top

AKG London: Erich Lessing 9bl; **Bridgeman Art Library, London / New York:** Musée Conde, Chantilly, France (MS 65/1284) / Giraudon 8; **Corbis:** Yann Arthus-Bertrand 41; Dean Conger 22; Les Gibbon / Cordaiy Photo Library 48; Tim Hawkins / Eye Ubiquitous 45tr; David Paterson / Wild Country 33; **Environmental Images:** M Bond 32bl; R Grace 37; C Jones 44tr; S Morgan 57bl; **Galaxy Picture Library:** SVS / GSFC / NASA 7t; **Robert Harding Picture Library:** 39bc; Robert Estall 40; J Greenberg 39tr; Gavin Hellier 39tc; **Nigel Lacey:** 29b, 43bl(bones), 65tr(inset); **Popperfoto:** Reuters 59br; **Science Photo Library:** 14bl, 68/69, 70/71; George Bernard 14br; Martin Bond 57c; Chris Butler 55; Bernhard Edmaier 5; Richard Folwell 64; Simon Fraser 34; Jeff Lepore 44bl; Tom McHugh 46; John Mead 47; NASA 31; NOAO 13; Novosti Press Agency 30tl; Dr Juerg Paren 10cl, 11; David Parker 42; Catherine Pouedras 52; Tom Van Sant / Geosphere Project, Santa Monica 2; O Tolstikhin / JVZ 6; **Scott Stickland:** 25.